CREATURES THAT EAT PEOPLE

RICHARD FREEMAN

CREATURES THAT EAT PEOPLE

WHY WILD ANIMALS MIGHT EAT YOU

Coral Gables

Published by Mango Publishing, a division of Mango Publishing Group, Inc.

Cover Design | Layout & Design: Megan Werner
Cover Illustration: C Design Studio, tiena / stock.adobe.com

For permission requests, please contact the publisher at:
Mango Publishing Group
2850 S Douglas Road, 2nd Floor
Coral Gables, FL 33134 USA
info@mango.bz

For special orders, quantity sales, course adoptions and corporate sales, please email the publisher at sales@mango.bz. For trade and wholesale sales, please contact Ingram Publisher Services at customer.service@ingramcontent.com or +1.800.509.4887.

Creatures that Eat People: Why Wild Animals Might Eat You

Library of Congress Cataloging-in-Publication number: 2023938920
ISBN: (pb) 978-1-68481-371-1 (hc) 978-1-68481-372-8 (e) 978-1-68481-373-5
BISAC category code: SPO030000, SPORTS & RECREATION / Outdoor Skills

Printed in the United States of America

For Pierre Denys de Montfort (1766–1820), the prophet of the Kraken, a man much misjudged and abused by history.

TABLE OF CONTENTS

◇◇◇

INTRODUCTION

Lust bares his teeth and whines
For he's picked up
the scent of virtue
And he knows the panic signs.

—"DIANA" BY COMUS

In 2021, a skeleton was unearthed in the Tsukumo burial site, a prehistoric hunter-gatherer cemetery in Japan's Okayama prefecture. The man was a hunter-gatherer-fisherman from the Jōmon period. The skeleton was radiocarbon-dated to 1370–1010 BC. The bones showed 790 gouges, cuts, and punctures, all made whilst the man was still alive and concentrated mainly on the arms, legs, chest and stomach. University of Oxford researchers J. Alyssa White and Rick Schulting concluded that the man had been killed by a shark, either a white shark (*Carcharodon carcharias*) or a tiger shark (*Galeocerdo cuvier*). Scans of the bones showed that most of the victim's ribs were fractured and bitten, and that his chest cavity and abdomen were probably eviscerated. The wounds were also concentrated on his left hip and leg, and he may have lost his left hand while trying to protect his body from the attack. At three thousand years old, this is the oldest evidence of a shark attack upon a human being.

Most of us have little to fear from large predators today. We live in urban environments where big, flesh-eating animals are long gone. Many countries, like the UK, have driven all their large, potentially dangerous meat-eaters to extinction, but it wasn't always that way. In the past, simple day-to-day living was a struggle for existence. Simply to make it from one day to the next was a triumph.

Several million years ago, our australopithecine ancestors on the plains of East Africa had to struggle to survive. They were being preyed upon by, and were in competition with, various formidable creatures. Crocodiles and pythons ate them, as did big cats, African hunting dogs, and large birds of prey such as eagles.

Maybe at the same time they began to grasp their own mortality and have a concept of death and the fact that they too would one day die, something that could happen at the drop of a hat in such a deadly, predator-filled environment.

It has been postulated that much of our psychology and behaviour can be traced back to these ancient plains. The monsters that haunt our legends may well be based on distorted race memories of the African predators. Why is it that in most houses the bedrooms are upstairs? We feel safer sleeping up high because our ancestors slept in trees to avoid predation at night. Why is it that evil is associated with the colour black? At night, human vision is poor, but predators like leopards (*Panthera pardus*) and other night hunters see clearly in the dark. The night was full of monsters for our ancestors. Even after *Homo erectus* harnessed fire some one million years ago, the monsters still lurked beyond that wan ring of light, waiting to pounce on those who entered the dark.

It has also been suggested that pareidolia (the way that the human mind sees shapes in random patterns like seeing faces in clouds or flames) may be a holdover from these times. An ability to recognize the dangerous shape of a predator as it lurks in the underbrush is a great survival trait to pass on.

The fossil remains of *Australopithecus robustus* have been found with leopard bite marks in the skull. A skull of *Australopithecus africanus* has been found pierced by the talons of an African crowned eagle (*Stephanoaetus coronatus*), while others bear bite marks from crocodiles. Death came in all shapes and sizes.

Mankind survived by its wits. Developing weapons, taming fire, and, above all, cooperating in groups. We left Africa and spread all over the world, becoming the dominant species, but in doing so we lost our fear of the wild. Most of us only see dangerous wild animals behind bars at the zoo.

We have a false notion that the whole world has been tamed, explored and made safe. In fact, massive areas of our planet are still wilderness areas inhabited by predatory creatures. For those who still live in these areas, man-eaters are still a very real fear. Each year humans fall prey to flesh-eating animals, but we in what we call the "civilized world" only hear of a fraction of those cases. It is generally when somebody from the white, English-speaking world falls victim to one of these creatures that the story makes its way into our media. When somebody from "our world" ventures into the wild and is in the wrong place at the wrong time, we feel an echo of that old fear our ancestors knew so well, the fear that may have helped us survive, to run that bit harder, to jump and climb that bit higher.

In this book we will examine man-eaters, animals that target, kill, and eat humans. There is a big difference between man-killers and man-eaters. Elephants, rhinos, hippos, giraffes, water buffalo, and other big herbivores sometimes kill people, but they are not man-eaters (save for one very strange case from World War II). Likewise, the many species of deadly venomous snakes around the world kill between 81,000 and 138,000 people per year, but none of them are man-eaters. All venomous snakes are too small to swallow a person. Their bites are purely defensive. These creatures are not the focus of this work. We are looking at animals that look on us as food: creatures that kill us, not because they think we are a threat or are infringing on their territory, but because they *want* to eat us. When we are back in the wild places, we are back on the menu.

CHAPTER ONE

CROCODILIANS

"Being eaten by a crocodile
is just like going to sleep…
in a giant blender."

—HOMER SIMPSON

On the night of October 12, 2002, two bombs exploded in popular nightclubs on the Indonesian island of Bali. A device in a backpack was detonated in Paddy's Irish Bar and, shortly after, a car bomb was used to destroy the Sari Club. Both were situated in the Kuta Beach area. It was peak tourist season, and 202 people were killed. The Islamic militant group Jamaah Islamiyah was behind the atrocity, claiming that the bombings were in retaliation for support of the United States's war on terror and Australia's role in the liberation of East Timor.

German tourist Isabel von Jordan, twenty-four, narrowly escaped the explosion in the Sari Club, leaving shortly before the bomb was detonated. To get over the shock, she joined her sister in Australia, where she visited other people who were hurt in the blast. But the hand of fate has a far reach.

She and her sister went on to visit Kakadu National Park, where they decided to swim in a pool despite the signs warning that crocodiles lurked in the area. Englishman James Rothwell, who was with the girls, said, "I heard a girl scream and she went under the water. I thought at first it was people mucking around."

Isabel had been grabbed, dragged underwater, and killed by a twelve-foot crocodile, which swam away with her body. Park rangers found the crocodile eating her corpse the following day and managed to kill it with harpoons.

Of all large, predatory animals, crocodiles present by far the biggest danger to humans, with around a thousand human deaths attributed to them each year. There are twenty-five living species in the order Crocodilia, including true crocodiles, alligators, caimans, and the gharial. They range in size from the five-foot-long Cuvier's dwarf caiman (*Paleosuchus palpebrosus*) to the saltwater or Indo-Pacific crocodile (*Crocodylus porosus*), which may approach or even exceed thirty feet. The bite force of a large crocodile may be as much as 7,739 lbs, over three times that of a great white shark. This force, almost four tons, would be like having the weight of a large truck slamming into its prey, a truck studded with sixty to seventy deadly teeth.

Though a number of crocodilians have been known to kill and eat humans, only two species, the Nile crocodile (*Crocodylus niloticus*) and the saltwater crocodile (*Crocodylus porosus*), are classed as regular man-eaters. We will look at these two species first.

THE NILE CROCODILE

The second largest reptile in the world, the Nile crocodile is widely accepted to be the most aggressive and dangerous crocodile species. It is found throughout sub-Saharan Africa and Madagascar, wherever there is sufficient water.

Nile crocodiles have been venerated across the African continent. Sebek was the crocodile-headed god of the Egyptian pantheon. He controlled the flow of the river Nile and was associated with the fertility of the Nile delta. When the river flooded in September, the delta would have been filled with crocodiles. He was also known as the great devourer, perhaps in reference to the crocodile's phenomenally powerful digestive system. This was seen as the natural circle, the life, death, and rebirth of all things. He is generally portrayed as a humanoid with a crocodile's head.

Despite his predatory nature, Sebek was a benevolent god. The son of Neith, the first Egyptian god, Sebek was said to be one of only two gods who would endure forever, as other gods' powers would fluctuate. He was a patron of the Thirteenth Dynasty kings (1800-1650 BC), many of whom were called Sebekhotep ("Sebek is satisfied"). His cult flourished in the delta areas, such as Fayoum, Thebes, and Lake Moeris. In 1900, excavations at Tebtunis uncovered a temple dedicated to Sebek, one hundred feet long and decorated with scenes of adoration, offerings, rituals, and great processions in his honour.

Another centre of his worship was Crocodilopolis ("crocodile city"). Here sacred crocodiles believed to be avatars of Sebek were kept in special sacred pools. They were adorned with gold and hand-fed milk and honey cakes by priests. Perhaps it is here that the idea that dragons guarded gold had its genesis. The chief of these pampered giants was known as Petersuchos. The city was said to have been founded by King Menes, first of the pharaohs. The legend tells that he was set upon by wild dogs and fled into the waters of Lake Moeris. Here a crocodile offered to carry the king upon his back to the site that later became the city.

Crocodiles, like all sacred Egyptian animals, were mummified. Specimens from one-foot hatchlings to sixteen-foot adults have been found mummified. At Kom-Ombo, small specimens were found stacked in the thousands, and, at the Maabdha Caves in central Egypt, crocodile mummies were found stacked to a height of thirty feet.

Human sacrifice to sacred crocodiles has been practised in several parts of Africa. British missionary Rev. J. Rosco described a crocodile cult he stumbled upon in Lake Victoria, Uganda. The high priest would mimic the movements of the sacred beasts and call them out onto the lake shore. Here prisoners, with their arms and legs broken, would be staked out. The crocodiles emerged and devoured the hapless victims. It was believed that the fishermen of the tribe could then cross the lake safely.

On the island of Madagascar, similar beliefs called Tangem-voay were held. Here crocodiles were known as *voay* and credited with the ability to distinguish between right and wrong. It was said that they never attacked anyone with a clear conscience. Many times an accused, in the heat of a legal debate, would rise up and shout, "May the *voay* eat me if I have done what I am accused of," and make for the nearest river. In one case, a girl was said to be having an affair with a slave and was condemned to trial by *voay*. On a full moon night, she entered a river close to an island inhabited by *voay*, then submerged herself three times. She was not attacked, and her accuser was ordered to pay heavy indemnity.

The French naturalist Chapelier, who visited the island over 180 years ago, knew of a chief in the Bay of Antongil who regarded a gigantic crocodile that lived in a nearby pool as his ancestor. Each year, he sacrificed one young man and one young woman to the beast. Bedecked in jewellery, they were hung over the pond until the huge reptile came to devour them. The story is very reminiscent of the virgins staked out for dragons in medieval Europe.

Lest you should think that sacrificing humans to crocodiles occurs only in the most savage reaches of the Dark Continent, you might want to think on a recent discovery in Naples. Excavations for a new metro near the Maschio Angioino, a thirteenth-century fortress, have turned up the skeleton of a big crocodile, lending credence to an old legend. The fifteenth-century queen Giovanna is said to have fed her lovers to a giant crocodile imported from Egypt. The beast was kept in a huge pit connected to the sea. Alexandre Dumas recounted this story in his *History of the Bourbons in Naples*. The crocodile's reign of terror was brought to an end when it was fed a poisoned horse haunch. Drawings of the castle in the eighteenth century show an embalmed crocodile displayed near the gateway.

The largest Nile crocodile ever officially recorded was twenty-one feet two inches long and weighed in at 1.2 tons. The crocodile was killed by the Duke of Mecklenburg in Tanzania in 1905. However, this is almost certainly not the size limit for the species. The "official" maximum sizes of many animals examined in this book are simply the largest accepted by experts. The idea that the largest measured specimen represents the absolute maximum for a given species is naive and somewhat arrogant.

Many larger specimens have been reported, and even measured, by reliable witnesses. The renowned wildlife photographer Cherry Kearton (1871–1940) and his friend James Barns observed a twenty-seven-foot crocodile basking on a sandbank in the Semliki River in Uganda. The size was estimated against other crocodiles and nearby objects. Kearton was an expert camera man and unlikely to have made an error. Such a monster would be well over three tons in weight.

A twenty-six-foot specimen was claimed by a Captain Riddick, who is alleged to have shot it at Lake Kioga in Uganda, and another of similar size was killed on the Mbaka River (in what is now Tanzania) in 1903. This was recorded by the experienced field naturalist Hans Besser. At first, he mistook the reptile for a huge canoe half drawn out of the water.

It was twenty-four feet nine inches long, but part of the tail was missing (almost certainly bitten off by another crocodile when it was young). The whole length with the missing tail section would have been twenty-eight feet, and the monster would have tipped the scales at least 3.5 tons. The body was three feet six inches high and was 14.72 feet in girth. The skull was 4.48 feet long.

In 1954, explorer Guy de la Ruwiere saw a twenty-three-foot crocodile in the Maika marshes in the northeast Congo. The animal lifted its massive head out of the water several times. It caused a huge wave when it dived beneath the surface.

One man who is very adept at estimating size is Rupert Bunts. I interviewed him at his Devon home about his experience. Mr. Bunts had been a soldier in Rhodesia (now Zimbabwe) in the early 1970s, and one of his jobs was to intercept terrorists from neighbouring Zambia. The easiest way to tell if a man was indeed a terrorist was by his boots, Zambian boots being different from Rhodesian ones. On one occasion, a suspect ran into the water in the southern end of Lake Kariba in an attempt to swim away from the patrol. The ill-fated fellow was seized and bitten in two by an immense crocodile. Rupert and his companions opened fire on the giant reptile with high-powered SLR rifles. Once the titan lay still, they drew alongside in a boat. When it was dragged ashore and cut open, the luckless victim's legs were retrieved. He was indeed a Zambian.

I asked Mr. Bunts how large the crocodile was. To my amazement, he told me it was between twenty-five and thirty feet long. Mr. Bunts was sure of this, as he was used to estimating distance and size as part of his job. Unfortunately, none of the men knew the zoological importance of the specimen, and no photos were taken or samples kept. A thirty-foot crocodile would weigh around four tons!

The largest reported crocodiles on the African continent hail from that last great African frontier, the Congo rainforest. They are known to the Lingala and other Congolese people as "Mahamba." This lord of the jungle is said to reach a shocking fifty feet in length!

In the late nineteenth century, Belgian explorer John Reinhardt Werner reported sightings of giant crocodiles that lend some weight to the terrifying folk tales of the native population. Whilst travelling down the Congo on the *Aja*—a forty-two-foot steam launch—Werner stopped at a sandbank to shoot ducks. He shot one and had pursued others over a low ridge when he saw "...the biggest crocodile I have ever seen. Comparing him to the *Aja*, which lay in deep water some three hundred yards off, I reckoned him to be quite fifty feet long: whilst the centre of the saw-ridged back must have been some four feet off the ground where his belly rested."

Werner stupidly took another shot at the ducks—they had run out of meat on the ship—and alarmed the monster, which made off into the water. The creature was also witnessed by a native boy Werner had with him. Around three days later, Werner saw another vast specimen. The *Aja* had embedded itself in a sandbank when it was heaved up out of the water by something causing a commotion under the ship.

> ***I saw an enormous crocodile—longer I am certain than the Aja—rush across the bank and tumble into the deep water beyond. I never before saw such a large crocodile move so fast, and I had no time to get a shot at him. He must have heard us coming and was trying to make for the deep water on our side of the bank, when we ran into him and rammed him onto the sand. We struck him, moving at a rate of four miles per hour, but during the short time he was in view I could not see that he bore any marks of the collision!***

It would be as well now to pause and reflect on the dimensions of such a huge crocodile. A twenty-five-foot creature would be an awesome animal, in the two-to-three-ton weight bracket. A fifteen-metre or fifty-foot animal would be of a colossal weight. When an animal doubles its size, its weight increases eightfold. This is because length, breadth, depth, and height have all been doubled. If we take the very conservative estimate of two tons for the weight of a twenty-five-foot specimen, then an animal fifty feet in length would weigh in the region of fifteen tons—three times the weight of an average elephant!

If crocodiles of these dimensions do exist, then they are the largest macropredators on the planet. Most of the great whales are plankton-feeders, and even the toothed sperm whale feeds mainly on small fish and squid (the giant squid forms only 1 percent of its diet and weighs far less than the sperm whale in any case). Such a giant crocodilian would be surpassed only by the giant marine reptiles of the Mesozoic, and possibly the largest carnivorous dinosaurs. (Palaeontologist Gregory S. Paul postulates a maximum weight of twenty tons for the largest tyrannosaurs, and this seems confirmed by a recently excavated specimen.) A fifty-foot crocodile would also exceed the largest known fossil species, the forty-foot-long *Sarcosuchus imperator*. If they do indeed exist, there is no animal on earth that could possibly withstand an attack from one of these giant saurians. Are we dealing with super-sized Nile crocodiles, a vast, unknown crocodile species, or simply wild exaggerations and traveller's tall tales here?

The Nile crocodile has a vast life span. A male Nile crocodile terrorized the Okavango Delta in Botswana. The man-eater was captured alive by an elephant hunter called Sir Henry in 1903, when it was already an adult. It was named Henry after its captor. Henry resides in the Crocworld Conservation Centre in Scottburgh, KwaZulu-Natal; he weighs 1,100 pounds and measures close to sixteen feet in length. He has fathered at least seven hundred offspring. Henry now lives with a harem of six female crocodiles. He has been in captivity for 115 years and must have been at least fifteen years old (and possibly far older) when he was caught. Henry is still living and is at least 130 years old.

Nile crocodiles prey on a wide variety of animals. Adult fare usually consists of zebra and various antelopes. However, much more formidable prey is taken, including spotted hyena (*Crocus crocuta*), giraffe (*Giraffa camelopardalis*), African leopard (*Panthera pardus pardus*), adult

African lions (*Panthera leo*), cape water buffalo (*Syncerus caffer*), and in very rare cases, even black rhinoceros (*Diceros bicornis*) and common hippopotamus (*Hippopotamus amphibius*) have fallen prey to Nile crocodiles. Attacks on African elephants (*Loxodonta africana*) are not unknown, though no attack on an adult elephant has been successful as far as is known. Only the mighty white rhinoceros (*Ceratotherium simum*) seems exempt from attacks.

Humans too form part of their diet. Just how many people are killed in Africa by Nile crocodiles each year is unknown but is likely to be in the hundreds. They can approach utterly unseen and attack with blinding speed.

Lake Rudolph, in the Great Rift Valley, lies in northern Kenya with its north tip entering southern Ethiopia. It is famous for its large population of crocodiles. Arthur Henry Neumann was a Victorian explorer. In his book *Elephant Hunting in East Equatorial Africa,* he vividly described how his cook, Shebane, was grabbed and killed by a huge crocodile on the shore of the lake in 1897.

> ***Having bathed and dried myself, I was sitting on my chair, after pulling on my clothes, by the water's edge, lacing up my boots. The sun was just about to set, its level rays shining full upon us. Rendering inconspicuous from the water while preventing our seeing in that direction. Shebane had just gone a little way off along the brink and taken off his clothes to wash himself, a thing I'd never known him to do before with me; but my attention being taken up with what I was doing, I took no notice of him. I was still looking down when I heard a cry of alarm, and, raising my head, got a glimpse of the most ghastly sight I ever witnessed. There was the head of a huge crocodile out of the water, just swinging over towards the deep with poor***

Shebane in its awful jaws, held across the middle of his body like a fish in the beak of a heron. He had ceased to cry out, and with one horrible wriggle, a swirl and a splash, all disappeared. One could do nothing. It was over: Shebane was gone.

On April 13, 1966, six young Americans from the Peace Corps, two women and four men, visited Gambela in Ethiopia. Professional hunter Karl Luthy was in the same village, building a pontoon on the river Baro. Locals had warned the Americans not to swim in the river, despite the heat, due to crocodiles that had recently eaten a child and a woman. Luthy himself reiterated the danger involved in swimming in the river.

A splashing noise made Luthy look up from his work, and he saw that the visitors had foolishly ignored the warnings of the local people and were swimming in the Baro about 450 feet away. Luthy was annoyed that they had ignored his warnings. Later, five of the six emerged from the water, but one, Bill Olsen, remained in the river, standing on a rock, waist deep in the water. Luthy continued working, but when he looked up again a few moments later, Olsen was gone.

Instinctually I glanced around, a prickle of apprehension spreading over me. But he was nowhere to be seen and I never saw him alive again, although we were to meet face to face much later when I fished his head out of the croc's belly.

The other members of the Peace Corps thought Olsen was still in the river swimming and would soon join them on the bank. However, Luthy saw the croc surface some fifteen minutes later, holding the man's corpse in its jaws. Looking through the hunter's binoculars, the other members saw their friend in the reptile's clutches and went to pieces. Luthy's client, a Colonel Dow, came down to see what the commotion was about and wanted to try and shoot the crocodile there and then. But Luthy persuaded him not to, as the animal was in midstream and Olsen's body would be lost. He was convinced that the crocodile would haul out of the water to bask and bring its prey with it the next morning.

Sure enough, the villagers told them the next day the crocodile was on a sandbank eating its victim. The colonel stalked downstream on foot, then crossed the river by canoe. The croc was indeed feeding on the remains of Olsen. He shot the creature in the neck with his rifle and the beast lunged into the water, taking its breakfast with it. He pumped three more bullets into the crocodile, but it did not halt it.

Luthy, the Peace Corps members, and a number of villagers joined Dow in dugouts from the village. The beast emerged onto another sandbank and Dow was able to shoot it a fifth time, finally killing it. Luthy and the others dragged what remained of Olsen from its jaws. His legs were intact, but the rest of the body had been ripped to bits and swallowed; the head was crushed in the powerful jaws before being swallowed. The crocodile was only thirteen feet one inch in length, quite small compared with some other monster specimens. It just goes to show how powerful even a small crocodile can be.

Crocodiles can display surprising levels of intelligence. They will work cooperatively when dismembering large prey items, with some individuals bracing the carcass whilst others tear chunks of flesh from the body. They will also climb upon each other's backs in the water to form a wall of crocodiles several individuals high to corral fish. Others have been seen using fish they have killed as bait to attract larger prey. This shows a higher intelligence, to be able to hold off the instinct to instantly eat the fish to get bigger and more satisfying food items in the future.

Dr. L. S. B. Leakey, the famed palaeontologist, saw evidence of the Nile crocodile's cunning whilst working on the island of Rusinga in Lake Victoria in the early 1930s. Walls of long, thick, tightly packed stakes had been erected, making enclosures to keep crocodiles out and allow people to draw water from the lake and water livestock. One crocodile, however, had learned by observation to crawl up out of the water and lurk in the bushes behind the enclosed area. There it would take both livestock and human prey. The village chief had asked for Leakey's help. He managed to shoot the animal behind the head as it lay on a mudbank. When its stomach was sliced open, many bangles and necklaces were found inside it. Villagers recognized these as belonging to lost family members.

Amy Nicholls, an eighteen-year-old British gap-year student, was traveling in Kenya with fifteen other students in 2002. They had spent five weeks performing volunteer conservation work, including surveying elephant and lion populations, on the edge of the nearby Tsavo National Park. The group visited Lake Challa, on Kenya's southern border with Tanzania. Assured by staff at a safari lodge that it was safe to swim, Amy and two other girls, Eleanor Lancaster and Elspeth Harley, set off for the lake at twilight.

"Amy swam a stroke and turned round to face the bank. Then she suddenly started screaming. I held my hand out to her, but in a few seconds she disappeared," Elspeth said.

Amy resurfaced several meters farther out and screamed that a crocodile had got her by the feet, then she vanished for good.

Walking towards the nearest police post, Elspeth met the safari lodge's owner. "She said we must be careful because there were pythons and crocodiles in the lake," said Elspeth. "Nobody told us anything about it being dangerous until afterwards."

Three days later, fishermen found her body entangled in their nets. The crocodile had only eaten one of her arms, leading them to believe it was a modest-sized one. Even a fairly small crocodile can be deadly.

Jake Grieves-Cook, a trustee of the Kenya Wildlife Service commented on the matter. "To be honest, I'm not aware of any lake in Africa which is perfectly safe to swim in."

Crocodiles are not afraid to attack small boats, overturning them, snatching victims out of them, or crushing them in their huge jaws. Dr. Richard Root was the Chairman of Medicine at Yale, UC San Francisco, and University of Washington. He also launched the Infectious Disease division at the University of Pennsylvania in 1971.

Dr. Root, when he was sixty-eight, was in Botswana at the invitation of the Ministry of Health to help the government treat HIV. He was providing care for patients, teaching interns and medical officers, and working on outreach programmes.

He took some time off in 2006 to see the country's amazing wildlife. He and his wife, Rita, went canoeing on the Limpopo River. They took two canoes, and each had a guide. Suddenly, a fifteen-foot crocodile lunged out of the water and grabbed Dr. Root, pulling him into the water to devour him. What remained of him was found some days later. The crocodile was never captured.

Some individual Nile crocodiles are thought to have killed and eaten vast numbers of people over the years. One that lived on the Kilhange in Central Africa devoured four hundred people over the years, until it was shot in the 1960s by hunter Pete Wessels. It was fifteen feet three inches long. Wessels was told of another croc on the Zambezi that had accounted for three hundred victims. A missionary teacher told him that it had taken seven of his pupils.

Between 1991 and 2005, a 16.4-foot crocodile killed eighty-three people near the village of Luganga in Uganda: that's 10 percent of the village population. The crocodile was named "Osama" after terrorist Osama bin Laden. He was reckoned to be around seventy-five years old and would not only snatch people from the shores of Lake Albert, where he made his home, but overturn and leap onto boats to attack prey.

In March 2005, a group of wildlife officials and fifty-odd locals captured the deadly animal, using cow's lungs as bait, with a snare. It took fifty men with ropes to haul Osama out of the lake. Taken by a company called Uganda Crocs, he now is part of a breeding programme and happily feeds on chicken rather than human flesh.

But the most infamous African man-eater, and possibly the largest, may still be at large. Christened "Gustave," he was made known by French environmentalist Patrice Faye. His home is on the Burundi side of the Rusizi Delta. Gustave is estimated to be between eighty and a hundred years old and may have eaten more humans than any other individual crocodile alive. Between 2003 and 2004, he is known to have eaten seventeen people. Locals say he has been dining on man flesh for over thirty years, so his human victim tally must be mind-boggling.

Locals exaggerated Gustave's size through fear, saying he was forty feet long. Faye estimated him to be over twenty feet. Having seen a photograph of him next to a full-grown hippo, I would put his length between twenty-three and twenty-five feet.

People had tried to kill Gustave before, including by shooting him with a machine gun. All this did was put a few dents in his hide.

A park ranger reported seeing the giant crocodile attack and kill a full-grown hippo by ripping its throat out. It is thought that Gustave hunts larger prey due to his massive size.

Patrice and his colleagues attempted to capture Gustave alive and put him on display as a tourist attraction to boost the finances of the Rusizi Game Reserve, which is being regenerated. Patrice had a thirty-foot cage constructed and used meat as bait, but Gustave was too canny. Baited snares were used and caught some small crocodiles, but not the giant.

In 2019, it was reported that Gustave had finally been killed, but no evidence of this has ever emerged.

THE INDO-PACIFIC OR SALTWATER CROCODILE

Officially, the world's largest reptile is the Indo-Pacific crocodile (*Crocodylus porosus*). This animal is also known as the "saltwater" or "estuarine" crocodile. It ranges from Northern Australia up through New Guinea and the surrounding islands and through Southeast Asia as far north as northern India. In former times it inhabited southern China, where it was known as the "flood dragon," possibly due to it coming into contact with humans during periods of flooding.

The Indo-Pacific crocodile feeds on a wide range of animals throughout its range, including deer, antelopes, water buffalo, gaur (*Bos gaurus*, the largest species of wild cattle), wild pigs, leopards, tigers, and even sharks. There is one account of a twenty-foot Indo-Pacific crocodile killing and eating an eighteen-foot great white shark, though details are lacking. Experts have often wondered which of these two great predators would prevail in a fight. My money is firmly on the crocodile, with its far more powerful bite and its ability to fight with large prey for long periods. In contrast, the great white has a tactic of bite and retreat; it is not made for prolonged battle.

Like its African cousin, much folklore has grown up around the animal.

On the Philippine island of Luzon, a particularly large crocodile haunted the mouth of the Cagayan River. It was believed to contain the soul of a dead chief. This man was apparently the leader of a fierce mountain tribe. Ergo the croc was left well alone.

In Indonesia, crocodile folklore is rampant. On many islands in this vast chain, it was thought that women could mate with crocodiles. The product of this union was a human baby and a crocodile. The little reptile was released into the river whilst the human child was taken home. The mother left food for her scaly child by the waterside. Its human brother or sister would carry this on as they grew up. In return, the crocodile would protect the family. On certain feast days, they would throw special foods to the crocodiles.

On the island of Buru, if a crocodile was terrorizing a community, it was believed that the offending creature had become infatuated with a local girl. Some poor woman was chosen (by what means is unclear, but it was probably akin to European witch hunts), dressed in a bridal costume, and given to the crocodile. There are no prizes for guessing what the wedding feast consisted of!

On the Moluccas, crocodiles played a part in puberty rites. Youths would be taken from their mothers and passed through the jaws of a replica crocodile. Then the priests would then take them away and teach them tribal secrets. After several days, they were passed out through the crocodile's jaws again and rejoined the tribe to great rejoicing.

Of all places in Asia, the island of Borneo has the most folklore concerning crocodiles.

Many of the legends tell of marriage to crocodiles or crocodiles protecting certain tribes or peoples.

One such story relates how Sarani, a princess of the Iban tribe, fell in love with a handsome prince. After a brief courtship, they were married, and the prince took Sarani away to his kingdom. They paddled in a canoe for several miles. The prince told Sarani to close her eyes. When she opened them, she was in a wondrous underwater kingdom, surrounded by fish.

After several months, Sarani told the prince that she missed her family and wanted to return. The prince said that if she left, she would never be able to return. After some thought, she said that she wished to be with her family again. The prince then revealed that he was the prince of the crocodiles. He gave her a parting gift, a magic jar that would bring prosperity to whoever owned it. If the jar was filled with water and it was sprinkled on the paddy fields, it would ensure a good harvest. Also, Sarani and all her descendants would be protected from crocodile attack.

The saltwater crocodile may have been behind some of the legends of dragons in the far East. In 1690, German explorer Engelbert Kaemfer was shown a dragon that was worshipped in a Japanese temple. He describes it thus: "A huge four footed snake, scaly all over the body like a crocodile with sharp prickles along the back; the head beyond the rest monstrous and terrible."

Was Kaemfer describing a crocodile? Crocodiles are not native to Japan, but the saltwater crocodile may occasionally stray outside of its northern range into the waters around some of Japan's most southerly islands.

Known for spending weeks at sea, these huge animals sometimes stray as far afield as New Zealand, where they may have generated the legend of the Māori dragon or *Kihawahine*.

Papuans along the Sepik River in New Guinea credit the crocodile as the great creator of all things. He caused the first dry land to appear from the primal waters. He formed a crack in the earth and mated with it. From that crack all animals and men came forth. When he opened his jaws, the upper jaw became the sky and the first dawn occurred.

The Itamul people there also tell that crocodiles roamed the new earth, founding villages. They carve crocodile heads on their canoes and statues of crocodile-headed men.

Crocodile skulls are often kept in men's cult houses and given offerings of betel nuts. During the initiation of Itamul youths into manhood, they are said to be swallowed by the primal crocodile, who spits them out as men. Their shoulders and torsos are subjected to ritual scarification. These, to the uninitiated, are the marks of the crocodile's teeth.

Crocodiles are powerful totems, and the consequences of breaking the totemic relationship can be fatal. A well-known folk tale involves Yali of Sor, leader of the Madang cargo cult. Whilst in the jungle, a comrade of Yali's killed a crocodile. Without the protection of the totem, Yali's friend became lost in the bush and died.

In Northern Australia, there is an Aboriginal legend of the crocodile's origin. A group of people were being transported across a river in a boat. An old man was waiting to be taken across. The boat came and went, picking up more people, but ignoring the old man. Eventually, he became so angry he leapt into the water and turned into a crocodile. Since then, crocodiles have attacked boats.

The Gunwinggu people of Arnhem Land believe that the Liverpool River bed was chewed out of the land by a giant crocodile who rose up inland, behind the mountains, and proceeded to munch his way out to sea.

The largest specimen generally accepted by experts was a twenty-eight-foot four-inch male shot on the MacArthur Bank of the Norman River, Queensland, in 1957 by Mrs. Kris Pawloski. The mammoth body was too big to move, but was photographed. Sadly, the photograph was lost in 1968. However, her husband Ron was a recognized expert on crocodiles, and had carefully measured the specimen. He was astounded at its size, having previously measured 10,287 specimens and found none larger than eighteen feet. This crocodile would have weighed around three tons.

Again, even larger specimens have been claimed. One of these was witnessed in the 1950s by rubber plantation owner James Montgomery. Montgomery's plantation was near the Segama River in North Borneo. The local Seluka tribe spoke of a monster that haunted the river. They called it "the Father of the Devil" and threw silver coins into the water to appease it.

Investigating, Montgomery found the beast hauled out on a sandbank. It filled the whole bank, with the end of its tail in the water. Wisely deciding to leave the monster well alone, Montgomery retreated. Returning later, he found that the sandbank it had been basking on was thirty feet across, indicating that the creature must have been more than thirty-three feetlong.

The great naturalist Charles Gould was well aware of giant crocodiles. A friend of his, Mr. Dennys, a resident of Singapore, told him of a thirty-foot crocodile that haunted a tidal creek that ran through the city in the 1880s. Another colleague—Mr. Gregory, the Surveyor General of Queensland—informed him that Australia's northern rivers were home to crocodiles as long as a whaleboat (twenty-eight feet).

There is a long history of giant crocodiles in Australia. In 1860, a thirty-three-footer was said to have been shot on the bank of the Mossman River.

Another of the same size was shot by publican Jack O'Brian from the veranda of the Leichardt Hotel on the banks of the Pioneer River. The beast was dubbed the Mackay Monster after the town the river ran through. The creature was put on display at the hotel and the skin kept there for some years, but it subsequently vanished.

Percy Trezise, pilot, artist, author, and expert on Aboriginal rock art, saw a giant crocodile whilst flying over the Cape York Peninsula with a group of tourists.

I was flying low over an estuarine area, with the tide out, and I came around a corner at 100 feet, and there was the crocodile of all crocodiles. It looked as if you couldn't reach across his back with spread arms. In my opinion he was more than 30 feet long.

Lest you think such giants are restricted to the past, another monster croc is reported in modern times. It is said to be over twenty-eight feet, and it inhabits the Guider River swamps of northern Arnhem Land.

The Guinness Book of Records attributed the greatest number of fatalities in an animal attack to the events on Ramree Island, Burma, on the night of February 14, 1945. It was the tail end of WWII, and British soldiers had forced a thousand Japanese troops into retreat on the island's mangrove swamps. During the night, the British heard splashing and the wild screams of the Japanese as huge Indo-Pacific crocodiles began to attack, kill, and eat them. Out of the thousand men who entered the swamps, only twenty made it out alive, the others meeting their fate in the jaws of the crocodiles.

Lieutenant-General Jack Jacob recorded the events in his memoir, *An Odyssey in War and Peace*.

Over a thousand soldiers of the Japanese garrison retreated into the crocodile-infested mangrove swamps. We went in with boats and interpreters using loudhailers, asking them to come out. Not a single one did. Salt-water crocodiles, some of them well over 20 ft (6.1 m) long, frequented these waters. It is not difficult to imagine what happened to the Japanese who took refuge in the mangroves!

As with the Nile crocodile, certain huge man-eaters are given names and live in infamy. One of the largest dwelt in the Dhamra River, located in the Bhitarkanika Sanctuary of Odisha, India. Named "Kalia," the giant beast haunted the river for over fifty years, preying on humans, particularly women doing their washing in its waters. Kalia was shot in 1926 by the captain of a ship going from Chandbali to Calcutta. The injured reptile crawled onto the bank, taking shelter in the reeds and dry grass. Seizing the opportunity, the villagers set fire to the vegetation, killing him. His body was measured and found to be between twenty-three and twenty-four feet long. The skull was preserved and kept by the rajah of Kanika. It welcomed visitors to the palace in Rajkanika, while the bangles and anklets found in his belly were displayed on a table.

In 2006, herpetologist Romulus Whitaker, founder of the Madras Crocodile Bank (an Indian conservation centre), measured the skull. He had worked out a ratio of skull to body length in crocodiles from measuring hundreds of them. The length of the Odisha man-eater was just over twenty-three feet. Bhitarkanika Sanctuary today plays host to another crocodile of the same size, dubbed "Mahisasur," meaning "big demon."

The island of Borneo is a hotspot for man-eating crocodiles. Margaret Brook was the wife of Sir Charles Brooke, the second white rajah of Sarawak. In her 1913 book *My Life in Sarawak*, she writes of a huge crocodile that haunted a local river.

> ***A great many years ago, before Kuching became as civilized as it is now, and when it had few steamers on the river, an enormous crocodile, some twenty feet in length, was the terror of the neighbourhood for three or four months during the north-east monsoon—the rainy season of the country. Our Malay quartermaster on board the* Heartsease *was seized by this monster as he was leaving the Rajah's yacht to go to his house, a few yards from the bank, in his little canoe. It was at night that the crocodile seized him, the canoe being***

found empty the next morning. Although no one had actually witnessed the calamity, it was certain the poor man had been taken by the monster. This was his first victim, but others followed in quick succession.

The crocodile could be seen patrolling the river daily, but it is very difficult to catch or shoot such a creature. At length the Rajah, becoming anxious at the turn affairs were taking, issued a proclamation offering a handsome reward to any one who should succeed in catching the crocodile. This proclamation was made with as much importance as possible. The executioner, Subu, bearing the Sarawak flag, was given a large boat, manned by twenty paddles, painted in the Sarawak colours, and sent up and down the river reading the proclamation at the landing-stages of Malay houses.

Looking from my window one morning, I saw the boat gaily decorated and looking very important on the river, with the yellow umbrella of office folded inside and the proclamation from the Rajah being read. A few yards behind the boat I imagined I could see, through my opera glasses, the water disturbed by some huge body following it. The natives had noticed this too, and it was absolutely proved that wherever the boat went up or down the river, the monster followed it, as if in derision of the proclamation.

The best-known Bornean man-eater was christened Bujang Senang, the Happy Bachelor. His story began on June 26, 1982. Bangan Anak Pali, an Iban farmer, and his brother Kebir had decided to collect shrimps in the Batang Lumpar river. Bangan had recently received a letter confirming his appointment as a chieftain of the Iban living around the Tanjung Bijat area.

Whilst wading in the river, Batang suddenly felt that he was standing on a large log. Then the water exploded. The man was swept off his feet by the crocodile's huge tail and seized in its huge jaws. Kebir said it was longer than their boat and had a distinctive white back.

A day later, the crocodile was seen flipping Batang's body about like a rag doll. Bujang Senang devoured most of it. Only the head and upper torso were retrieved.

In another river, the Tebu, an eighty-six-year-old Iban man, Abang Anak Gelayah, was taken by a massive crocodile that rammed and overturned his boat. His daughter-in-law and granddaughter managed to swim to the bank, but the old man was grabbed and dragged under.

Soon after, the local police launched Operation Buaya Ganas (Ferocious Crocodile). It was led by Sarawak Commissioner of Police Datuk Seri Yuen Leng and involved men armed with modified bear traps, harpoons, guns, and grenades. A group of medicine men from Indonesia joined the hunt. One of these claimed to have harpooned Bujang Senang from a boat, but he was so powerful that he dragged the vessel against the current, almost sinking it, and broke free, dislodging the harpoon.

The crocodile took bait, such as monkeys, from hooks, but never got caught. At one point, ducks with grenades attached were employed! Bujang Senang declined such delicacies. Hunters threw spears at the monster, but they simply bounced off his scales. On another occasion, two grenades were hurled into the water, but the killer croc just sank and vanished.

It was two years before Bujang Senang struck again. On September 27, 1984, a fifty-one-year-old Iban man, Badong Anak Apong, was fishing for shrimps close to the site of the first attack. The crocodile rammed his boat and capsized it. It grabbed Badong and dragged him to a muddy bank whilst thrashing him violently. Five men on the riverbank witnessed it.

Running to the tribal longhouse, they alerted six hunters who returned to the river with rifles to see the huge crocodile shaking and tossing the man's body. Despite shooting it twice, they were unable to get it to let go. The victim was devoured. Again the distinctive white back was reported. A large reward was offered for anybody able to kill or catch Bujang Senang, but nobody claimed the cash. All was quiet for the next few years.

Five years later, on February 20, 1989, farmer Berain Anak Tunggling was repairing his boat in shallow water when the white-backed crocodile dragged him off his boat and into the water.

A seventy-year-old man named Tuah Anak Tunchun came forward and said that the same white-backed crocodile had eaten his brother Inchi in 1962 at Sungei Selumbang. Others claimed to have seen it in the company of an even larger black (in reality very dark green) crocodile. It was postulated that five other people devoured between 1965 and 1979 were victims of Bujang Senang.

On May 21, 1992, thirty-year-old Dayang Anak Bayang was wading across a tributary to hunt. Unknown to her, Bujang Senang was lying in wait on the bed of the tributary. Her mother, Umain, saw Dayang suddenly vanish under the water. Then a huge crocodile broke the surface, holding the woman in its jaws. Umain flailed at the monster with a tree branch to no avail. With a flick of his tail, the beast carried Dayang away into the water.

Several people who were nearby heard the commotion and ran to help. Saperie Anak Kiang shot at the beast but missed. Dayang's brother, Enie Anak Bayan, was fishing downriver and was informed of the attack. Armed with a shotgun, he paddled upriver and saw the crocodile with his sister in its mouth. He shot it twice, to no effect.

News reached the longhouse where Dayang lived, and by noon, twenty-six people with guns and spears had gathered at a deep pool where Bujang Senang was thought to be lurking. Enie found his sister's body in the reeds, and it was retrieved and put on a boat. Umain fainted at the sight.

Suddenly, the crocodile emerged to reclaim its prey. Enie, from only ten feet away, fired his shotgun, hitting it in the eye. It thrashed violently and sank. Mandauu Anak Tabor, a neighbour who had joined the hunt, managed to throw a spear into the monster's back. The crocodile made its way upriver with the spear still jutting from its back. It smashed through a fence across the river. And it was shot again and again. Bujang Senang continued upriver until he found his way blocked by a fallen tree, and swam back to the pool. Mandauu bravely tried to force the spear deep into the beast's back, but it bent on the iron-hard scales.

Jaws agape, the crocodile lunged forward as Mandauu, Enie, and another man, Sidi Anak Iman, shot into its mouth at point-blank range. The killer reared up out of the water, standing on its tail. Crashing back down, it bit madly at logs and debris in the water before sinking its teeth into a tree stump and expiring.

It took twenty-four men to haul the massive body onto a boat. The Happy Bachelor was measured at nineteen feet, three inches, and tipped the scales at over a ton. He was a big crocodile, but by no means the biggest.

The "black" crocodile sometimes seen with Bujang Senang was even bigger, according to witnesses. Pandi Anak Lai said, "I have seen a black crocodile which is at least 35 feet long, along the Batang Lumpar."

Bujang Senang may have bitten the dust, but there were other killer crocodiles stalking the rivers of Borneo.

Not only natives fall prey to the jaws of Indo-Pacific crocodiles there. In 2002, Richard Shadwell, a thirty-five-year-old musician from Sutton, in Surrey, was eaten by a crocodile in the Simpangkanan River in Tanjung Puting National Park. In an act of extreme foolishness, Richard was swimming behind a boat in the crocodile-infested river.

"Shortly after Mr. Shadwell plunged into the river, I saw a black crocodile devour him, then he vanished," a guide called Jeki told the Indonesian News Agency.

In 2022, the *Daily Star* reported the capture of a twenty-six-foot man-eating crocodile in the Semaja River in North Kalimantan on Borneo.

The crocodile had killed and eaten Samsul Bahri, forty-five, as he fished for shrimp on July 19.

Police and villagers captured and sedated two crocodiles fourteen and sixteen feet long. The crocodiles were induced to vomit, but no human remains were found in their stomachs. Later, the twenty-six-foot giant was snared. Again, this crocodile was sedated and induced to vomit. The contents of the creature's stomach contained partially digested human remains that were identified as those of Samsul. No further details of this case have come to light at the time of writing.

Moving on to Australia, there have been predatory attacks in the tropical north of the country.

In September of 1920, the *HMS Geranium* dropped anchor in Vansittart Bay, in the Kimberly region of northwest Australia. Gunner H. Davies and his fellow sailors were allowed some shore leave. Davies was engaged to be married once his ship reached Fremantle. His fiancée was already on a passenger liner en route to meet him there.

Whilst his peers decided to go fishing, Davies elected to explore the area. Unfortunately, he got lost, and as darkness fell, he did not reappear. Aboriginal trackers were used to search for the man. They found tracks showing boots worn from the harsh terrain and showing dragging, exhausted steps. The tracks led to a creek, then vanished.

On October 9, his leg was found. On October 15, his body was found minus the head, one foot, and both arms. Davies had fallen prey to a crocodile. His remains were buried at a mission in Pago. The stone cross with brass lettering still stands in the abandoned graveyard.

In 1948, British refrigeration engineer John Thompson vanished from a freighter tied up at a wharf in Wyndam. No trace of him was ever found. Nearby meat works discharged blood into the river on a regular basis, attracting large numbers of crocodiles.

Tex Boneham, a crocodile hunter, killed an eighteen-foot croc and found a gold ring inscribed with the letters JT in its stomach. It was sent back to Britain and identified as belonging to John Thompson.

On July 1, 1980, Bukarra Munyarrwun, an Aboriginal woman, was fishing in a billabong near the isolated Dhalinby settlement in Northern Territory. She was in an area where the Cato River joins Arnhem Bay. She was a mother of two and led a traditional Aboriginal life, subsisting by hunting and fishing.

Friends heard her scream and saw an eighteen-foot crocodile grab her. Police were called and came in a four-wheel drive along the bumpy road from the town of Nhulunbuy, thirty-one miles away. Police Sergeant Bob Haydon found Munyarrwun's torso; her lower half had been eaten.

The Aborigines refused to allow the crocodile to be shot. They believed he was an ancestor spirit from the dreaming that they called "Baru."

The people instead performed a series of "corrobrees," ritual dances to appease the woman's spirit and plead with their totem beast not to kill anybody else. As far as is known, Baru still resides in the billabong and has never killed anybody else. Nobody knows how big he has grown in the forty-three years since.

On Sunday, December 21, 2003, three friends, Ashley McGough, nineteen, Brett Mann, twenty-two, and Shaun Blowers, nineteen, came to a spot near the Kangaroo Flats, Northern Territory, to race their quad bikes. At four thirty in the afternoon, they went down to the Finnis River to wash off. But they had not noticed how fast the water was flowing.

Shaun later recounted what happened.

The three of us walked into the water among some stringy trees. The water was running a little bit at this spot...and Brett went out a little farther and was washed away. I don't know if he lost his footing or the current was a bit strong for him. After we saw Brett washed away, both Ashley and I went out after him. Ashley and I caught up to Brett and we both got in front of him as we went with the flow. I was in front, Ashley was next, and Brett was at the rear. We were all within arm's reach of

each other. It probably took us about 300 metres to catch up with Brett, and then we began to look for a place to get out of the river. We all spoke to each other to check that we were all right. There was no real panic at this stage.

Caught up in the current, they were swept along for almost a mile.

Ashley yelled out, "Croc, croc, I'm not joking, there's a fucking croc. Head for a tree, get out of the water." I didn't see a croc but swam to the nearest tree and climbed up into the first fork. I helped pull Ashley up into the same tree. We looked around for Brett and called his name out. I didn't see Brett anywhere or hear him call out. I didn't hear a call or a splash or anything. It wasn't very long after we got into the tree, maybe two minutes later, that I saw a croc pop up with Brett in his jaws. Brett wasn't moving, he was lying face down in the water and the croc was gripping him by the left shoulder. I know it was Brett because he was wearing his O'Neill riding gear, which was mainly yellow with black and white stripes. The croc was only about five metres away from us at the time. It was only a couple of minutes that the croc remained looking around at us. It went under the water with Brett and swam away. I did not see Brett again.

The crocodile was about thirteen feet long and highly aggressive. The two friends scrambled into forks in the tree to escape it. It lurked by the tree all night watching them, and they scrambled even higher. At one point Shaun fell into the water but managed to scramble out and up the tree again. They spent the whole night clinging to the branches, terrified of the predator below. They clung to each other for warmth and to keep each other awake. Shaun recalled:

> ***Whenever we moved, we'd say, "I'm moving," and just check in on each other and make sure we weren't going to sleep. We were worn out from hanging on to the little tree. The tree was swaying all night because there was a lot of wind and rain.***

When they did not return home, the alarm was raised and Marine and Fisheries Enforcement Section and the Territory Response Section made their way to the area with family friend Wayne Mitchell. The group spotted the boys, but the crocodile was still prowling the waters below.

A helicopter usually used on an oil rig was dispatched. It has a six-man crew—pilot Captain Wayne Silby, senior first officer Max Neill-Gordon, winch operator Gordon McRae, wireman Milton Ellis, paramedic Michael McKay, and police sergeant Glenn McPhee. However, the boys could not be airlifted out for fear of the downdraft from the blades knocking them into the water.

Men were lowered onto a small island in the river and inflated a life raft, which they paddled over to the tree, acutely aware of how flimsy the craft was against a crocodile's jaws. The boys had to jump down from the tree into the raft.

The noise of the helicopter apparently scared the crocodile away. No trace of its victim was ever found.

THE MUGGER CROCODILE

Crocodylus palustris, the mugger crocodile, is found in fresh water across most of India and Sri Lanka. It is also found in parts of Pakistan, Iran, and

Nepal. It became extinct in Bhutan in the 1960s but was reintroduced into the Manas River in the 1990s.

A veneration of crocodiles, similar to that of ancient Egypt, once existed in India and Pakistan. The Victorian traveller Andrew Leith Adams wrote of it after a visit in the 1860s. The "Mugger Peer" lived in a sacred pond in Karachi, which in those days was part of India. The pond contained many crocodiles, but the "Mugger Peer" was preeminent among them. A huge specimen, he was reputedly two hundred years old and had his head painted red. He was attended by priests who fed him and his kin. Travellers were expected to make donations to both the priests and the crocodiles. Adams and company had a goat sacrificed to them and watched as the dismembered beast was fed to the holy crocodiles.

The mugger pool still exists today, together with its population of holy reptiles. It is a Sufi shrine. This liberal strand of Islam has no problem with the veneration of the muggers. It is known as the Mango Pir shrine and lies around fifteen miles southwest of the city centre.

Muggers feed on fish, birds, and small mammals, but as they grow larger, they take larger mammals, like deer, antelope, and water buffalo. On rare occasions, they have been known to kill leopards and tigers. In 2017, in southern Sri Lanka, a mugger attacked and seriously mauled a wild elephant!

The mugger has been observed to use tools. They balance sticks in their snouts and lure birds to perch on them or to try to use them for nesting material. When the bird lands on the stick, the crocodile grabs it.

Males are usually around eleven or twelve feet long, but on occasion can grow to eighteen feet. They are not nearly as aggressive as Nile or Indo-Pacific crocodiles, but they do sometimes prey on humans.

In 2017, Paul McClean, twenty-four, a British journalist for the *Financial Times*, was killed by a mugger crocodile at a lagoon in the coastal village of Panama, southeast Sri Lanka. McClean, who was on holiday, had gone to the toilet, then down to the water to wash his hands, when a crocodile bit into his leg and dragged him off.

Other holidaymakers alerted police and a search was mounted with the help of navy divers. His body was found with deep puncture wounds on the leg.

In July 2020, a mugger crocodile attacked and killed a fifty-four-year-old woman from Mahadevpura village in Waghodia Taluka, who was washing clothes on the banks of the Dev River in Vadodara district, Gujarat, India.

It dragged her away by a leg, a major portion of which it bit off during an hour-long fight with villagers trying to rescue her.

Hemant Vadhana and his team of crocodile catchers joined the forest department team led by Kashyap Patel. The crocodile was lassoed and pulled from the river to be secured and relocated.

This creature had killed seven people over a seven-year period. Officials say it may have turned into a man-eater because medical waste is often found in the river near a cancer hospital about a mile away.

THE AMERICAN ALLIGATOR

Alligator mississipiensis is found throughout the warm southeast of the US. There are a number of differences between true crocodiles and alligators. The largest crocodiles are far bigger than the largest alligators. Crocodiles generally have longer, more triangular jaws and alligators shorter, more rounded ones. The crocodile's bite is stronger, though the alligator's jaws are far more powerful than any shark or big cat. When the crocodile's jaws are shut, both upper and lower teeth are visible, whereas it is mostly only the upper teeth of the alligator that are seen. Crocodiles have salt glands that allow them to tolerate sea water; alligators lack these. But the main difference lies in the integumentary sense organs. These sense movement in the water and appear like black dots on the animal's scales. Alligators have them only along the jawline, but crocodiles have them all over the body.

The American alligator may reach as much as fifteen feet long, but most are smaller. There is one story of a nineteen-foot, two-inch alligator shot and measured in 1890 by Edward Avey Mclhenny in Louisiana. However, Mclhenny was a known teller of tall tales and saw himself as somewhat of an entertainer. Unlike Nile and Indo-Pacific crocodiles, there has only been one such claim of an outsized individual.

One of the earliest accounts dates from 1836, when the King family was crossing the Trinity River in Texas. Their ferry ran aground, and the head of the family, Gray B. King, jumped into the water to push the boat free, but was torn to shreds and eaten by alligators.

In May 2006, Yovy Suarez Jimenez, a twenty-eight-year-old biology student, was out jogging near a suburban canal in Sunrise, a dormitory community north of Fort Lauderdale, Florida,. She was attacked and partially eaten by a ten-foot alligator. The Broward County medical examiner, Joshua Pepper, told reporters, "The alligator attacked her and basically amputated her arms, bit her on the leg and back and pulled her into the water. She died extremely fast. By the time she was pulled into the water, she was already dead."

A witness said she had seen Jimenez dangling her legs into the water. High levels of alcohol and Xanax, an anxiolytic, were found in her system.

The role that human stupidity plays in animal attacks should never be underestimated. On May 1, 2020, despite the Covid lockdown, Cynthia Covert visited a friend to do her nails and drink wine with her. After finishing her work, she saw an alligator in a pond at the gated community on Kiawah Island, near Charleston, South Carolina, where her client resided. Covert approached the creature and tried to pet it like a dog. Her friend warned her that she had recently seen the alligator kill a deer. Covert replied, "Don't look like a deer."

The alligator bit into her and dragged her under. Her friend tried to save Covert by throwing her a rope. She dragged the victim out of the water, but the alligator attacked again, biting into Covert's leg, hauling her back underwater, and killing her. Her last words were, "I guess I won't do that again."

In 2007, police responded to a call about people breaking into vehicles in the parking lot of the Miccosukee Indian Reservation resort and convention centre in Dade County, Florida. Miami resident Justo Antonio Padron, thirty-six, was one of two men police found there and chased. Padron had been wanted by authorities since September for violating probation after pleading guilty in June to cocaine possession. Officers arrested his accomplice, Heriberto Rubio, thirty-seven, but they say Padron ran from them and jumped into the lake behind the resort. His body was found the next day, killed by a nine-foot, three-inch alligator.

OTHERS

Other crocodilians can reach large sizes, but seldom feed on human prey. The false gharial (*Tomistoma schlegelii*) averages about fourteen feet, but a skull in the Munich Museum suggests an individual of twenty feet. It feeds mainly on fish and thus has evolved narrow jaws. It takes birds and mammals up to the size of cows. In 2008, a fourteen-foot female attacked and ate a fisherman in central Kalimantan. His remains were later found in the stomach. By 2012, two more fatal attacks on humans were recorded.

The Cuban crocodile (*Crocodylus rhombifer*) is found only in Cuba in two regions: the Zapata Swamp and Isla de la Juventud, an island off the south coast. It can reach sixteen feet, but today, individuals over eleven feet are rare. This species is highly aggressive and territorial. It is thought that they once fed on giant ground sloths and other now-extinct megafauna, and this is why they are so aggressive. Some of their teeth are more curved than those of other crocodiles, and they are known to spend more time on land than other species. There is some evidence of pack hunting in Cuban crocodiles. These are all adaptations for hunting large land mammals. Surprisingly, only one known human death is attributable to this species, possibly due to their very limited range. If they were more widespread, there would probably be more human fatalities attributed to the Cuban crocodile.

In South America, there are several large crocodilians, but again, they do not seem to target humans as prey. The Orinoco crocodile (*Crocodylus intermedius*) can reach twenty-three feet, though individuals this large are apparently unknown today; these were in the nineteenth century.

Orinocos prey on fish, birds, deer, anacondas, caimans, monkeys, deer, capybaras, peccaries, jaguars, and tapirs. On his trip to the region in 1800, natives told explorer Alexander von Humboldt that two or three adults were killed by them per year. In recent years, there have been no attacks, possibly due to low population numbers. The species is now critically endangered, found only in Colombia and Venezuela, with a small group in Trinidad. Only 250 to 1,500 remain in the wild, with a small population in zoos.

The black caiman (*Melanosuchus niger*) is the largest of the alligator family, reaching twenty feet. It is found in Bolivia, Brazil, Colombia, Ecuador, Venezuela, French Guiana, Guyana, Peru, and Suriname. It feeds on fish (including piranhas), birds, and mammals, like monkeys, deer, capybaras, peccaries, jaguars, and tapirs. It also eats other, smaller caimans and anacondas. Despite its size and power, attacks on humans are rare. Explorer Henry Walter Bates noted a case in his 1863 book *The Naturalist on the River Amazon.* Whilst staying at Caicarra, Venezuela, an elderly justice of the peace saw a drunken man wading into a muddy river where a big black caiman was known to lurk. Despite his shouted warnings, the man continued to wade until waist deep. Then a massive black caiman exploded from the water, seized him, and dragged him away.

CAUSES OF PREDATORY CROCODILE ATTACKS

The interesting question of why some species of crocodilian are more inclined to prey on humans than others may be down to evolution. The two most dangerous species, the Nile and Indo-Pacific, have evolved alongside large mammals and still live alongside them, in most of their range. These species have evolved to hunt large mammalian prey. Humans are large mammals, and if you are on the riverbank in their habitat, you are on the menu. Species like the American and Orinoco live in areas where large mammals are absent or rare; therefore, the instinct to hunt and kill humans is not as acute.

Human behaviour is another factor in fatal crocodile attacks. Joachim John Monteiro was a British mining engineer, colonial administrator, naturalist, and explorer. He saw several instances of this during his time in Angola. On one occasion, a hunter had shot a hippo, and local people were cutting it up for meat. The blood was running down into the water and attracting crocodiles. Despite this, other people were wading into the water between the dead hippo and the lurking beasts to fill water containers.

In another instance, a woman was seen washing her baby on the shores of Lake Albert as a big crocodile swam by. When a warning was shouted to her, she replied, "Crocodiles do not eat people." Attacks were seen as providential. Crocodiles, the locals thought, only attacked bad people. Somebody with a clear conscience could enter a crocodile-infested river with impunity!

Naturalist C. A. W. Guggisberg noted similar attitudes. An African once told him that if a crocodile killed a person, then never mind, it was God's will. At Lake Baringo, Kenya, locals told him that the crocodiles at Lake Victoria were man-eaters, but those at Baringo were harmless.

In the case of the American alligator, the recent increase in attacks is doubtless due to human encroachment. In times past, they were rare, but now, as people live cheek by jowl with alligators, more are bound to occur. When you build condos and swimming pools in alligator territory, you are simply asking for trouble.

HOW TO AVOID ATTACK BY CROCODILES

If you are in an area where crocodiles live, then do not bathe, wash, or fish in the water. Never dangle legs or arms from boats, and never gut fish or prepare meat near the shore. If you are camping, do so far away from water. Both Nile and Indo-Pacific crocodiles have been known to venture onto land and attack people in camps, particularly at night.

If the worst happens and a crocodile sinks its teeth into you, then attack its most vulnerable spot, the eyes. During WWII, American entomologist P. J. Darlington was collecting mosquito larvae in New Guinea when a crocodile grabbed his arm. He managed to free himself by jabbing it in the eye.

James Kerr, a doctor, politician, senator, soldier, surveyor, and congressman, saved one of his Black slaves who had been sleeping outside his tent in southern Texas. The man was attacked by an alligator, and Kerr got the beast to release him and retreat by jabbing its eye with a burning branch from the campfire.

How effective this method would be on a really big croc is debatable. Sometimes big crocodiles kill with a single, crushing bite. When a human is up against a creature that can kill tigers, sharks, and water buffalo, then there is little chance of survival. There is only one true defence from the attack of a big crocodile, and that is reincarnation!

CHAPTER TWO

CONSTRICTING SNAKES AND KOMODO DRAGONS

"Kaa was everything that the monkeys feared in the jungle, for none of them knew the limits of his power, none of them could look him in the face, and none had ever come alive out of his hug."

—RUDYARD KIPLING,
"KAI'S HUNTING," *THE JUNGLE BOOK*

Jean-Claude Savoie owned an exotic pet store called Reptile Ocean in New Brunswick, Canada. He would often have the sons of his best friend, Mandy Trecartin, come in to play with his own son and see the animals. In August of 2013, six-year-old Connor and four-year-old Noah Barthe were sleeping in a room above the shop. Upon checking in on them in the morning, Savoie found both boys dead. Coiled nearby was a fourteen-foot African rock python.

It appeared that the animal had escaped from its enclosure in the collection below and slithered into the ventilation ducts. It then crawled up into the ceiling above the boys. It broke a hole in the ceiling, fell down on the two boys, and constricted them where they lay. A pathologist who carried out postmortem examinations said both boys had died of asphyxiation and had bite marks on their bodies.

Jean-Claude Savoie was later cleared of criminal negligence.

The only snakes that prey upon man are the big constrictors, the pythons and boas. They are the only snakes large enough to overpower and eat a human, and they have been doing so since our ancestors came down from the trees.

Pythons and boas kill not by crushing, as commonly assumed, but by constriction, a form of suffocation. The snake's muscular coils tighten when its victim breathes out, preventing it breathing in again, so it asphyxiates. Like all snakes, they swallow prey whole, employing extreme cranial kinesis or skull flexibility. The snake's lower jaw is not one solid bone, but two. The halves are connected by an elastic ligament. Thus the jaw can flare out to accommodate prey larger than the snake's head. The skin, tissue, and ligaments also stretch. The backwards-pointing teeth grip, and as the two halves of the bottom jaw can move independently, they can drag the prey into the throat by moving like a pair of hands pulling at a rope, the teeth providing the grip. Waves of muscle contractions then force the prey into the stomach. Prey is swallowed headfirst to help streamline it.

Pythons and boas are not closely related, despite resemblance and similar lifestyles. For the most part, pythons inhabit the Old World—

tropical Africa, Asia, and Australasia—whereas the boas live in the neotropics of the New World. The largest snake in the fossil record is *Titanoboa cerrejonensis*, which reached forty-seven feet and hunted the swamps of South America sixty to fifty-eight million years ago. The major physical difference between pythons and boas is that pythons lay eggs and boas give birth to live young. Despite being the largest of snakes, only a handful are predators of man.

THE RETICULATED PYTHON

The largest living python, *Malayopython reticulatus*, is a creature of the tropical rainforests of Southeast Asia. The species was immortalized by the great English writer Rudyard Kipling in his books *The Jungle Book* (1894) and *The Second Jungle Book* (1895) as Kaa. The original books are a million miles from the infantile bastardization created by Disney in the odious animated film. Kaa is not the cowardly second-string villain, but one of the oldest and wisest creatures of the jungle, and he fears nothing.

As with most big predators, there is much folklore attached to this creature. The name "python" itself derives from a giant serpent in Greek mythology. It was the child of Gaia, the earth goddess, and guarded the Delphic Oracle, the most important shrine in all Greece, which dated back to 1400 BC, when the god Apollo slew Python with magic arrows.

In his 1900 book *Malay Magic*, Walter William Skeat writes of strange beliefs on the Malayan peninsula with regard to pythons and their gall bladders.

> ***The gall-bladder of the python,*** *uler sawah*, ***is in great request among native practitioners. This serpent is supposed to have two of these organs, one of which is called*** *ampedu idup*, ***or the live gall-bladder. It is believed that if a python is killed and this organ is cut out and kept, it will develop into a serpent of***

> ***just twice the size of that from which it was taken. The natives positively assert that the python attains a length of sixty to seventy feet, and that it has been known to have killed and eaten a rhinoceros.***
>
> ***In Selangor, as in Perak, the "live gall-bladder" of the python will (it is believed), if kept in a jar, develop into a serpent; when dried it is in great request as a remedy for small-pox.***

Whilst I was searching for the orang-pendek, an unknown species of ape, in Sumatra, I was told of tribal spirits called "naneks." People on the island, though nominally Muslim, still retain old animist beliefs from centuries past. The jungle tribes each have a *nanek*, and one of these tribes has a *python* nanek. These are thought of as ancestor spirits who each founded the tribe in question.

The American animal collector and filmmaker Frank Buck heard of similar beliefs in the Philippines. In 1932, he was helping a native man look for his missing son. Eventually they noticed a large bulge in the belly of a twenty-five-foot reticulated python that the man kept at his house. It transpired that the snake had eaten the boy. Instead of grieving, the man told Buck he had "gone to sleep inside a god."

The largest reticulated python on record was thirty-three feet long, caught in the Celebes (now Sulawesi), Indonesia, in 1902. There are stories of much bigger specimens.

The eighteenth-century explorer Francis Legaut claimed to have encountered one fifty feet long on Java. A brute of similar size was reported in the *North China Daily News* of November 10, 1880. The story tells of a Western hunter (whose name is never revealed) who came across a remote hut in the dense jungle between Buddoh and Sirangoon on the Malay peninsula. Upon the roof was the skin of a gigantic python. Inquiring as to its origin with the hut's owner, he was told this story:

The Malay was awakened one night by his wife's screams. Investigating, he found to his horror an immense snake that had drawn the poor woman's whole arm into its maw and was in the process of swallowing her. The plucky fellow seized two bags and stuffed them into the corners of the giant reptile's mouth, thus forcing them to open wider. The snake released the woman and turned upon the man, whipping its coils about him. Fortunately for the Malay, his arms were free, and he grabbed his parang and hacked at the vasty serpent. The snake unwound and slithered through an opening beneath the hut. Both the man and the hut were covered in blood.

Come morning, he followed the python's trail to a patch of plantain palms. In its death throes the beast had smashed the trees and uprooted them. In the midst of the destruction lay the offending creature, dead. He had been offered 60 dollars from some Chinese who had travelled long distances to buy pieces of the monster's flesh due to its medicinal properties. They also offered him 6 dollars for the skin, this he kept however, as a trophy of the ordeal. The skin was between 7 and 8 fathoms (50–56 feet) long.

The huge size and apparent ease of the serpent's dispatch raise some scepticism in this case.

Another dramatic story is recounted in the Victorian natural history tome, *The Pictorial Museum of Animated Nature*:

The captain of a country ship, while passing the Sunderbunds, sent a boat into one of the creeks to obtain some fresh fruits, which are cultivated by the few miserable inhabitants of this inhospitable region. Having reached the shore, the crew moored the boat under a bank, and left one of their party to take care of her.

During their absence, the lascar who remained in charge of the boat, overcome by the heat, lay down under the seats and fell asleep. Whilst he was in this happy state of unconsciousness an enormous boa (python) emerged from the jungle, reached the boat and had already coiled its huge body round the sleeper, and was in the very act of crushing him to death, when his companions fortunately returned at this auspicious moment, and attacking the monster severed a portion of its tail, which so disabled it that it no longer retained the power of doing any mischief. The snake was then easily dispatched, and was found to measure, as stated 62 feet and some inches in length.

If this event actually occurred, the creature would have been an outsized reticulated python, not a boa. Once again, the ease of its killing raises suspicion.

Reticulated pythons feed on mammals such as monkeys and deer. They are also recorded to have killed and eaten sun bears (*Helarctos malayanus*) and leopards (*Panthera pardus*). In a staged fight, filmed for the 1932 film *Bring 'Em Back Alive*, a reticulated python fights a full-grown tiger (*Panthera tigris tigris*). The big cat is very nearly constricted to death and only just manages to pull itself from the python's coils and flee.

Of all the big constrictors, the reticulated python attacks humans the most often. As far back as 1638, Antonio van Diemen, Governor-General of the Dutch East India Company, recorded that a twenty-three-foot python had killed and swallowed a slave woman working in a garden on the island of Banda Api, Indonesia. The python was shot by Dutch soldiers and the Governor-General saw the huge snake with the victim still inside it.

The *Bombay Courier* of August 31, 1799, carried the following story.

A Malay prow was making for the port of Amboyna; but the pilot, finding she could not enter it before dark, brought her to anchor for the night, close under the island of Celebes. One of the crew went on shore in quest of betel-nuts in the woods, and on his return lay down, as it is supposed, to sleep on the beach. In the course of the night, he was heard by his comrades to scream out for assistance. They immediately went on shore; but it was too late, for an immense snake of this species had crushed him to death. The attention of the monster being entirely occupied by his prey, the people went boldly up to it, cut off its head, and took both it and the body of the man on board their boat. The snake had seized the poor fellow by the right wrist, where the marks of the fangs were very distinct; and the mangled corpse bore evident signs of being crushed by the monster's twisting itself round the neck, head, breast, and thigh. The length of the snake was about thirty feet; its thickness equal to that of a moderate-sized man; and on extending its jaws, they were found wide enough to admit at once a body of the size of a man's head."

In 1927, in Burma, the Viennese herpetologist, zoologist, and explorer Franz Werner (1867–1939) recorded the death of a man in the coils of a reticulated python. The jeweller Maung Chit Chine and some friends, while they were hunting in Thaton district, were caught in a violent hurricane. The jeweller took refuge in a tree, away from his companions. He never came down again. In the end, his friends only found his hat and shoes next to a python six meters long that had swallowed something. They killed the snake and cut it open. Inside they found the body of Chin.

Ecologist Harry W. Greene and anthropologist Thomas N. Headland made a study of the Philippine Agta Negritos when they were still preliterate hunter-gatherers. The Negritos were once widespread in the Philippines and elsewhere in Southeast Asia; however, by 1990, transition to a sedentary peasant lifestyle was complete, and now they are threatened with extinction. The study began in 1962, at Casiguran, in the Sierra Madre of Aurora Province, Luzon, when they were preliterate, lived in small kin-related groups, slept in tiny temporary shelters, foraged in old-growth rainforest, and ate wild meat daily.

They recorded six fatal attacks by pythons on the tribespeople between 1934 and 1973. They also found that 26 percent of adult males had survived predation attempts by reticulated pythons. More men were attacked then women, as the men hunted in the forests and encountered the pythons. Many who survived attacks had substantial scars from the python's teeth. The men defended themselves with bolo knives (akin to machetes) and homemade shotguns.

In 2018, on Muna Island, Southeast Sulawesi, a fifty-four-year-old woman, Wa Tiba, went missing while checking her vegetable garden. Village chief Faris said that a search party went out to look for her by torchlight, but found only her belongings, including sandals.

Later, villagers found a twenty-three-foot python 150 feet from her garden with a large bulge in its belly. They killed the snake and carried it to the village.

"When they cut open the snake's belly, they found Tiba's body still intact with all her clothes," Faris said. "She was swallowed first from her head."

The victim's garden, about a half-mile from her house, is in a rocky area with caves and cliffs believed to contain many pythons.

In October of 2022, a fifty-two-year-old woman named Jahrah went out to work on a rubber plantation in the province of Jambi, Sumatra, Indonesia. Her husband became worried when she did not return and reported her missing.

Setting out to look for her, he discovered her sandals, headscarf, jacket, and the tools she used at work, and called for others to help, police told local media. The next morning, a twenty-three-foot-long python with a large bulge in its belly was found.

Local police chief AKP S Harefa said, "When the security team and residents conducted a search around the rubber plantation, then we found a python 7 metres long. It is this snake that is suspected of preying on the victim. After we caught him, we found the victim's body in the snake's stomach."

The village residents were scared, as other huge pythons lurked on the plantation and had recently eaten two goats. A week before the attack, three villagers had caught a twenty-six-foot python on the plantation, but it overpowered them and escaped.

THE AFRICAN ROCK PYTHON

The largest snake in Africa, the African rock python (*Python sebae*), is found throughout sub-Saharan Africa. Python worship and python legends are widespread throughout their range.

The San people of northwestern Botswana believe that mankind descended from the python, and the ancient, arid streambeds around the hills are said to have been created by the python as it circled the hills in its ceaseless search for water. In 2006, associate professor Sheila Coulson of the University of Oslo discovered evidence of python worship in Botswana dating back 70,000 years. Coulson made the discovery while searching for artefacts from the Middle Stone Age in a group of small peaks within the Kalahari Desert known as the Tsodilo Hills, famous for having one of the largest concentrations of rock paintings in the world.

The Tsodilo Hills are still a sacred place for the San, who call them the "Mountains of the God" and the "Rock that Whispers." Professor Coulson's find showed that people from the area had a specific ritual location associated with the python. The ritual was held in a little cave on the northern side of the hills. The cave itself is so secluded and access to it is so difficult that it was not even discovered by archaeologists until the 1990s.

The cave was twenty feet long by six and a half feet wide. The opening resembled a snake's mouth. The rock itself was covered in three to four hundred indentations that could only have been man-made. The rock and cave had been altered to look like a python. Coulson wrote, "You could see the mouth and eyes of the snake. It looked like a real python. The play of sunlight over the indentations gave them the appearance of snake skin. At night, the firelight gave one the feeling that the snake was actually moving."

Excavating beside the rock, her team found more than 13,000 artefacts. All were spearheads and articles that could be of ritual use, as well as tools used in carving the stone. Some were 70,000 years old. Some were brought from hundreds of miles away. The spearheads were better crafted and more colourful than other spearheads from the same time and area.

> ***Stone age people took these colourful spearheads, brought them to the cave, and finished carving them there. Only the red spearheads were burned. It was a ritual destruction of artefacts. There was no sign of normal habitation. No ordinary tools were found at the site. Our find means that humans were***

more organised and had the capacity for abstract thinking at a much earlier point in history than we have previously assumed. All of the indications suggest that Tsodilo has been known to mankind for almost 100,000 years as a very special place in the pre-historic landscape.

The Luo people of Kenya and Tanzania worship pythons and call them "n'ilo." They believe them to be a reincarnation of the goddess Omieri, who is linked with rainfall, fertility, and the harvest.

The Fipa people of southwest Tanzania believe that the python can be both good and evil. Some are revered forest spirits and others are dragon-like monsters. One such serpent was known as "Zimwi," meaning chaos. It shot fire from its nostrils and uttered awful roars. It stole and ate all the game the first men killed. Zimwi was finally slain by Matanji the toolmaker, a Fipa culture hero.

The African rock python is the largest snake in Africa. One shot in Gambia measured twenty-four feet seven inches. Another twenty-four-footer was killed in Adiopodume in the Ivory Coast. The reader will not be surprised that far larger rock pythons have been reported.

An African rock python of thirty-two feet was supposedly shot near Bingerville, in the Ivory Coast.

The Victorian lady explorer Mary Kingsley spent many years in Africa and heard of giant pythons.

I am assured by the missionaries at Calabar that there was a python brought into Creek Town in the Rev. Mr. Goldi's time that extended the whole length of the Creek Town mission-house veranda and to spare. The python must have been over forty feet. I have not a shadow of doubt it was. Stay-at-home people will always discredit great measurements, but

experienced bushmen do not, and after all, if it amuses stay-at-homes to do so, by all means let them; they will have dull lives of it and it don't hurt you, for you know how exceedingly difficult it is to preserve really big things to bring home, and how, half the time, they fall into the hands of people who would not bother their heads to preserve them in a rotting climate like West Africa. The largest python skin I ever measured was a damaged one, which was twenty-six feet.

A monster snake was photographed by the passenger of a Belgian military helicopter pilot in the Katanga region of Zaire in 1959. The reptile is pictured so clearly that even the scales on its hide are visible. The photographer was one Colonel Remy van Lierde, an accomplished and decorated WWII pilot, working as a charter pilot at the time. His chopper crossed a hollow in a jungle clearing and he saw a vast snake emerging from a hole. The huge serpent was dark green with a lighter underside. Van Lierde, who was adept at estimating size, put it at fifty feet long. He made several passes, allowing his passenger to shoot a picture. The snake reared up fully ten feet, as if to strike at the helicopter. Van Lierde estimated the head at three feet long by two feet wide and likened it to the head of a gigantic horse.

African rock pythons eat a wide variety of animals, including birds, antelopes, monkeys, warthogs, hyenas, and leopards. One twenty-three-foot specimen was killed by K. H. Kroft in 1958 and was claimed to have had a four-foot, eleven-inch juvenile Nile crocodile in its stomach. They can and do take humans from time to time.

In 2002, a ten-year-old boy was killed and swallowed over a three-hour period by a twenty-foot rock python. The boy's three friends, who had been picking mangos, took refuge up a tree as the snake constricted and swallowed their friend near Lamontville, Durban, South Africa. One of the boys, Khaye Buthelezi, said:

We all kept quiet because we didn't want the snake to know we were there too. The snake squeezed tighter and tighter around him until his eyes closed and his head fell back so I thought he was dead or had fainted. Then the snake's mouth opened very, very wide and started to swallow him from the head down—his clothes and everything. It all took about three hours because it was dark when we saw it slither away and we finally came down from the tree.

Khaye led journalists back to the spot where the attack happened. He said that he had been having nightmares about being eaten by a python himself.

Snake experts and the police followed the python's trail from the scene to a nearby stream the snake had used to get away to digest its prey. It was never captured.

THE GREEN ANACONDA

It is in South America that we find the largest of all living snakes, the green anaconda (*Eunectes murinus*). It lives east of the Andes in rivers, lakes, and swamps. Adults spend almost all their time in the water. It ranges from Colombia and the Guianas in the north to Paraguay in the south, and all countries in between where there is sufficient water.

According to the observation of the chronicler Garcilaso de la Vega, in 1609, Amazonian Indians worshipped the great snakes because of their ferocity and thought themselves descended from the serpents.

The Desano, an Amazonian ethnic group that lives in the upper basin of the Vaupes River, believed that a great ancestral anaconda penetrated the universe through a water gate and ascended the Negro and Vaupes rivers carrying within its body the ancestors of all mankind. Along their journey, the remote ancestors transformed into humans. Upon reaching the Ipanore waterfall, formed by the flow of the Jirijirimo River, which in the indigenous language means "the bed of the anaconda," the ancestors divided the land into what became their territories.

In some areas, the sacred anaconda is called Yacumama or Sachamama, the mother of the waters and mother of the earth respectively. She is the protector and originator of life. To enter heaven, the dead had to seduce her. Those entering the water blew a conch to protect themselves from her.

The Huaorani people of Ecuador equate the anaconda to the Milky Way. They think that as the anaconda descended to earth, it created the Amazon River.

Father Pierre Charlevoix, a Jesuit missionary (1681–1761), wrote of a cult worshipping a giant anaconda in his book *The History of Paraguay*, published in 1769, after his death.

Alvarez, during his attempt to reach Peru from Paraguay, is reported to have seen:

> ***...the temple and residence of a monstrous serpent, whom the inhabitants had chosen for their divinity, and fed with human flesh. He was as thick as an ox and seven-and-twenty feet long, with a very large head and very fierce though small eyes. His jaws, when extended, displayed two ranks of crooked fangs. The whole body, except the tail, which was smooth, was covered with round scales of a great thickness. The Spaniards, though they could not be persuaded by the Indians that this monster delivered oracles, were exceedingly terrified at first sight of him, and their terror was greatly increased when on one of them having fired a blunderbuss at him he gave a***

roar like that of a lion, and with a stroke of his tail shook the whole tower.

We should note here that snakes do not roar, and a twenty-seven-foot-long anaconda would not be as thick as an ox. It would need to be over forty feet to attain that girth, unless it was gorged with a large prey item.

The largest specimen seems to be the one measured in Brazil in 1962 by W. L. Schurz, at twenty-seven feet nine inches, with a maximum girth of three feet eight inches. But more than any other large snake, the green anaconda has amassed sightings of giant individuals.

Damon, a chief of the Eagle Clan Arawak Amerindians, told that, at a remote pool known as Corona Falls, a gigantic anaconda had been seen. He had spoken with the hunters who had seen the beast. They told him that it was so large they had fled from it. When he asked how big the snake was, one of the men pointed to a thirty-foot palm tree. He told Damon that a dead tree of the same size had been lying in the water. The anaconda was crawling over it and its head and tail extended beyond the ends of the tree. This would make the snake around forty feet long.

In 1944, another large specimen was encountered in Colombia by a team of prospecting geologists led by Roberto Lamon. The men shot the snake and measured it at thirty-seven feet six inches. The group left the creature to eat their lunch, intending to come back and photograph their trophy and skin it. Upon their return, they were amazed to find it gone. The bullets had merely stunned the animal, which had recovered and absconded in their absence.

Federico Medem—a Colombian herpetologist—saw an anaconda that he estimated to be between thirty and forty feet and obtained a report of another thirty-four feet long. He apparently also saw a thirty-three-foot, eight-inch snake killed on the Guaviare River.

General Candido Mariano de Silva Rondon—who lent his name to the Rondonia area of Brazil—saw a specimen, killed by Indians, some thirty-eight feet long. There are several records of snakes in this size bracket that cannot easily be dismissed, as some have involved reputable scientists. A thirty-four-foot anaconda was shot by Vincent Roth, director of the National Museum, in British Guiana (now Guyana). Mr. R. Mole—a naturalist who made many important contributions to knowledge of the wildlife of Trinidad—reported a thirty-three-foot example there in 1924.

The Marquis de Wavrin was another explorer of South America and was active in the years before WWII. He told the great Belgian cryptozoologist Bernard Heuvelmans that he had seen anacondas over thirty feet long, and that the natives told of far larger ones. He once shot a twenty-six-foot individual that had been coiled around a branch. When he expressed a desire to retrieve the cadaver, his canoemen told him it was a waste of powder to shoot such a small snake and a waste of time picking it up.

They went on to say, "On the Rio Guaviare, during floods, chiefly in certain lagoons in the neighbourhood, and even near the confluence of this stream, we often see snakes that are more than double the size of the one you have just shot. They are often thicker than our canoe."

Colonel John Blashford-Snell, the legendary British explorer, was told a most intriguing story whilst travelling across the Andes by river from Bolivia to Bunenos. A forty-three-foot anaconda was captured by a farmer after it had eaten a cow. He apparently incited it with a pig on a rope. Subsequently, he tried to sell his story, unsuccessfully, to the press. The creature is now said to be residing in a pond on a farm in northwest Brazil. This occurred in late 1999.

Whilst taking part in the Scientific Exploration Society's Kota Mama IV expedition in 2002, Jerome Jennings of the Extreme Science website met Johnny, a soldier with the Bolivian Special Forces. He told Jennings the following story.

> ***I was working as a driver to a captain and had taken him to a meeting at the base near Riberalta. Very dangerous jungle, very wild, and had to wait for him. I knew that I had to hide if I did not want any more work so I climbed a tree at the edge of***

the camp and tried to relax. The tree was big and I could see over the jungle to a pool. The ground had been cleared to one side and the people who lived there had cows. A small cow, a baby cow, a calf went to take a drink at the edge of the water and a big anaconda came out of the water and takes the calf by the face and drags it into the pool. I nearly fell out of the tree when I saw this. When my captain had finished his meeting I told him and he ran into the office and grabbed a rifle. We went to the pool and shot the anaconda. The captain took the skin and put it in his house. It goes around the walls two times.

Jennings asked Johnny if the snake had been measured.

We did not measure it but it was more than 10 metres (33 ft). I know this because we saw another big snake outside the camp and it was not as big as the one we shot. The other snake went over the road and went past a low wall. It was night and I was in my jeep and saw the snake in the road but was too frightened to get out so I let it go. When it had gone I measured the wall. The wall was 10 metres and the snake has its head and tail past the wall as it went. The snake at the pool was much bigger.

Anacondas feed on birds, deer, monkeys, capybaras, tapirs, small caimans, and jaguars. Accounts of anacondas killing people are rare, but they do occur.

The late 1950s brought perhaps the most dramatic encounter with a man-eating anaconda. The political climate, with the resurgence of communism in Latin America, was such that the US government placed CIA agents in sensitive areas. One agent—called "Lee"—was told by a cattle rancher of a giant snake lairing in a cave in Bolivia. It was said to be over thirty-three feet long and to have eaten ten men and many cattle over the years. Every three months or so, it emerged, seized a steer, dragged it into the river, killed it, and ate it. Then it would return to its cave.

The rancher wanted Lee to capture the animal and take it to a zoo, as it was "probably the largest snake in the world." The problem was discussed at the embassy many times, until someone came up with an audacious plan to flush the monster from its lair with tear gas, whilst a long sack (with zip fasteners) was held over the cave's mouth. There would be two "zip-men"—one at each end of the sack—to hasten the operation. For added security, Lee carried (ironically) a .357 Python pistol.

It was just as well Lee was "packing heat," as things did go spectacularly wrong. The tear gas was shot into the cave, and the anaconda—thrashing madly—shot out of the cave, and into the sack. Once its entire length was inside, both ends were zipped up. The agents had not reckoned with the snake's vast strength, however. Its violent writhing split the sack—end to end—and the brute was free.

The livid animal came rushing at Lee, who whipped out his pistol and managed to put a bullet in its head. The snake threw itself into a huge loop, smashing into a small hardwood tree as big as a telephone pole. The tree was shattered like matchwood, and the snake fell back into the jungle. Lee pumped another two bullets into its head. When it had expired, they measured it at thirty-four feet seven inches. Lee skinned the snake and took the hide back to the US, where he kept it in his garage. Its current whereabouts are unknown.

Lee's colleague David Atlee Phillips understandably doubted his friend's outlandish story. Sometime later, he was attending a Washington party and mentioned the saga to Darwin Bell, then Deputy Assistant Secretary for International Labour Affairs. Bell claimed not only to have known Lee, but to have taken part in the attempt.

"I was the tail zipper man," he told an amazed Phillips.

THE KOMODO DRAGON

In 1912, the governor of the Indonesian island of Flores (then under Dutch rule), J. K. H. van Steyn van Hensbroek, led an expedition to investigate stories of a dangerous monster called the *buaya darat* on the island of Komodo. A year before, a lieutenant from the Netherlands had sent him photos of the creatures. The result was the discovery of the Komodo dragon (*Varanus komodoensis*), the largest known lizard in the world. The Komodo dragon lives only on a handful of Indonesian islands, including Komodo, Flores, Rinca, Padar, Gili Dasami, and Gili Montang.

Folklore about the creature abounds. It was said that the sultan of Sumbawa, a neighbouring island in the chain, would banish wrongdoers, rivals, and other undesirables to these islands to be devoured by the beasts. Pearl fishermen brought back hair-raising stories of encounters with the monsters. The beasts were said to have razor-sharp teeth and claws and a taste for human flesh.

Island natives say that when large Komodo dragons enter the sea, they engage in battle with crocodiles, and if defeated, return to the dry land and remain dragons. But if the dragons prove victorious, they stay in the sea and turn into crocodiles.

Selangor Malays claim that crocodile hatchlings that run in the direction of dry land are eaten by their mothers, while any that escape turn into Komodo dragons. This is probably based on the way that female crocodiles take their hatchlings down to the water in their mouths.

It is said that in ancient times, a dragon princess married a chief called Empu Najo. A pair of twins was born. One was a baby boy named Gerong, and the other a female Komodo dragon who was later named Ora. Gerong and Ora were reared separately. One day, when grown up, Gerong was hunting and met Ora, his twin sister. He unsheathed his weapon. However, suddenly, a mystical figure of his mother showed up. “Do not kill her; she is your sister,” The mother said.

Since then, the Komodo community have believed that they and Komodo dragons are siblings.

Komodo dragons can grow to over ten feet long and are the largest venomous creature, producing a hemotoxin that acts as an anticoagulant in the blood of victims. It kills with razor-sharp teeth and its venomous saliva, taking down deer, water buffalo, wild boar, and even humans.

Heru Rudiharto, head of administration at Komodo National Park, said that between 1974 and 2012, five people had been killed by Komodo dragons.

Dr. Walter Auffenberg, of the Florida State Museum, stayed on Komodo and Flores islands between 1969 and 1973 studying the creatures. He came across rogue individuals that were highly dangerous. One large male, which had been given the identity number 34W, had killed three people in the Poreng area of Komodo. The local people named it *ora gila*, meaning "crazy monitor." Auffenberg's own children were stalked by 34W. A tourist vanished in the area, leaving only a bloodstained shirt and camera. Auffenberg was certain that 34W had eaten him.

In 1974, an elderly Swiss tourist, Baron Rudolf von Reding Biberegg, fell and injured his knee on a hiking trip on Komodo Island. His guide returned to a village to seek help. All the search party found was the man's hat, camera, and a bloodstained shoe.

The predators are not afraid to enter villages. Traditional houses are built on stilts on the islands. In one attack, a girl was climbing down the ladder from her house when a Komodo dragon grabbed and killed her.

On Flores Island, a Komodo dragon snatched a ten-year-old boy from the veranda of a house. His mother struggled with the lizard and forced it to let go, but the boy was already dead.

In 2007, an eight-year-old boy called Mansyur was attacked and killed in scrubland in Komodo National Park. "The Komodo bit him on his waist and tossed him viciously from side to side. A fisherman, who just happened to be the boy's uncle, threw rocks at the lizard until it let the boy go and fled," national park spokesman Heru Rudiharto said.

The boy died from massive bleeding.

In New Guinea, there is another huge lizard, the crocodile monitor or Salvadori dragon (*Varanus salvadorii*). Though not nearly as bulky as the Komodo dragon, it is longer thanks to its elongated tail. It can reach fifteen feet. There are persistent stories of much larger lizards on the island.

In WWII, Japanese soldiers caught glimpses of what they described as "tree-climbing crocodiles" deep in the Papuan jungle. Then, in the summer of 1960, panic broke out on the island as rumours that people had been killed by twenty-foot-long dragons began to circulate. The monsters were said to breathe fire and drink blood. Their victims were left with foot-long claw-marks in the flesh. The scare became so bad that the government authorities moved people in the stricken areas into stockades and offered substantial rewards for the capture of one of the beasts. The reward went unclaimed, the dragons disappeared, and the riddle went unsolved for the next twenty years.

In 1960, Lindsay Green and Fred Kleckhan—two administration agricultural officers—found some skin and a jawbone of one of the dragons held as relics in a village near Kariuku. Today, they would be able to identify these specimens via DNA analysis, but such things were unheard of back then.

In 1969, David M. Davies, an explorer, was shown Papuan cave paintings of what looked like a giant lizard standing on its hind legs. His native companions reacted with fear to the picture. Late in 1978, a specimen was finally filmed in southern Papua by Jean Becker and Christian Meyer. However, even this could not determine if this was a new species.

In the mid-1980s, famed British explorer Colonel John Blashford-Snell was told of the "tree-climbing crocodile." Locals called it Artrellia and seemed to go in great fear of it. He was told that it stands upright and breathes fire. From the descriptions given to him by an old chief, he sketched an animal looking much like a dinosaur.

One story told of a young warrior who, many years ago, was hunting deep in the forest. Feeling weary, he sat down on a log. The "log" revealed itself as a huge lizard. It towered ten feet tall on its hind legs, and possessed toothy, crocodile-like jaws. The man fled back to his village in terror.

Intrigued, the colonel hit the trail. No less a man than the brother of the Premier of the Western Province told him that an elderly man had died in the Daru hospital after being attacked by a female Artrellia protecting her nest. A village elder also said the creatures could grow to over fifteen feet long, and often stood on their hind legs, lending them a dinosaur-like appearance. They were arboreal, and leapt down onto their prey, which they killed with their huge claws and infectious bite. Even small specimens were feared. A short time before, a small one had been captured and placed in a wooden cage. It swiftly broke free and killed a large dog before escaping back into the jungle.

The colonel searched for the dragon himself, but had no success. He then offered a cash reward for anyone who could bring him a specimen. Eventually, a village priest shot a small Artrellia. It was identified as the Salvadori dragon. The colonel later saw several twelve-foot specimens himself, and one huge individual with a head as large as a horse's was also seen. Such a vast specimen would be in excess of twenty-three feet.

Robert Grant and David George were exploring the Strachan Island district in 1961 when they encountered a grey-skinned lizard some twenty-six feet long. The creature's neck alone measured three feet.

In 1999, two groups of people spotted a dinosaur-like creature at Lake Murray near Boroko. It was twenty feet long, with crocodile-like skin. It had thick hind legs with smaller front limbs and a long tail. Both of these sightings seem to refer to gigantic Salvadori dragons.

In March 2004, a new sighting of a Papuan dragon emerged.

POLICE HUNT "DINOSAUR" IN PNG
MARCH 12, 2004—12:38 P.M.

Reports a live dinosaur had been sighted on a volcanic island of Papua New Guinea prompted the deployment of heavily-armed police in search of the mystery creature. Villagers in

the superstitious island province of East New Britain this week said they fled in terror after seeing a three-metre-tall, grey-coloured creature with a head like a dog and a tail like a crocodile. They said the creature was living among thick green plants in a mosquito-ridden marsh just outside the provincial capital Kokopo, near the devastated town of Rabaul, which was buried by a volcanic eruption in 1994.

Kokopo's mayor, Albert Buanga, said the dinosaur would make a great a tourist attraction, if it existed. A government official today confirmed police carrying M-16s and shotguns searched the area but found no trace of the creature. Eyewitness Christine Samei told reporters she ran for her life after seeing a three-metre-tall, grey creature with a head like a dog and a tail like a crocodile which was as fat as a 900-litre water tank. "It's a very huge and ugly looking animal," Samei told local media. A government official said the villagers had identified the creature from books and movies about dinosaurs. "They told us it was a dinosaur," the official told AAP. Although police found no trace of the creature, Senior Sergeant Leuth Nidung warned villagers to take extra precautions when going about their daily business, amid reports it had eaten three dogs. Villagers were told to report any further sightings immediately to police, who were already organising a more thorough search of the area.

Black magic and other superstitions are common in many parts of PNG's predominantly village-based society. Each year large numbers of foreigners visit the area to see World War II relics as well as the devastated town of Rabaul—the only urban centre in the world built inside the crater of a giant volcano.

◇◇◇◇◇◇◇◇◇◇◇◇◇◇◇◇◇◇◇◇◇

CAUSES OF ATTACKS BY CONSTRICTING SNAKES AND KOMODO DRAGONS

Humans are not the natural prey of constricting snakes. They feed mainly on four-footed mammals. Humans, being bipeds, present a larger outline that will often discourage pythons and anacondas from attacking. Our outward-facing scapulae or shoulder bones present a challenge for all but the largest snakes. Even with their flexible jaws, it is hard for most snakes to swallow humans, and only the largest can do it.

Most pythons and anacondas are ambush predators, waiting patiently for prey to come by. The reticulated python is a more active hunter, particularly at night. It is unlikely that constricting snakes deliberately target humans; they simply attack prey they think they can overpower, and with very big snakes, humans fall into that size range.

Komodo dragons kill with a venomous bite. They will tackle prey much larger than themselves, such as water buffalo. They evolved to prey on large mammals, such as the now-extinct stegodon elephants that once lived on Indonesian islands. A comparison could be made with the Cuban crocodile, another island-dwelling reptile that once preyed on now-extinct large mammals. It evolved specialized teeth, pack hunting, and an aggressive nature. The Komodo dragon developed venom to do the same job. As a hunter of large mammals, it will sometimes attack humans as prey. Like the Cuban crocodile, their restricted range stops them from killing many people.

HOW TO AVOID ATTACKS BY CONSTRICTING SNAKES AND KOMODO DRAGONS

If you are in an area where big constrictors are present, always travel in company—two, or preferably more, people should be together. This way, if one person is targeted in an attack, the others can help. In zoos, keepers try never to enter an enclosure with a big constrictor alone. Others can help free the victim from the snake's jaws and help unwind its coils.

If a snake bites you, do not try to pull away. The backwards-pointing teeth will rip further into your flesh. The best way to make a snake release you from a bite is to push your limb further into the animal's mouth. This sounds counterintuitive, but snakes never swallow live prey. A living prey item could thrash around inside the snake, causing internal damage. Snakes always kill prey before swallowing it. If you push an arm or leg further into the animal's jaws, it will free it from the teeth and cause the attacker to let go. However, if a big constrictor gets a lone human in a firm grip in several coils, there is little even the strongest person could do.

Komodo dragons only live on a handful of islands. Tourism is strictly controlled, so you are highly unlikely to be attacked by one, unless you are stupid enough to wander away from an official tour. Again, on these islands, travel in groups. If attacked by a Komodo dragon, fight back with all you have got. Use sticks and rocks, shout to others for help. The lizard may back down if faced with a long struggle from a determined victim.

CHAPTER THREE

SHARKS

"Oh, the shark, babe, has such teeth, dear, and it shows them pearly white."

—"MACK THE KNIFE," BOBBY DARIN

In 1940, Mrs. Burch and her two daughters Denise and Pamela were evacuated from the British colony of Hong Kong to Sydney, Australia. Her husband and son, who had been fighting in the war, were now prisoners of the Japanese, and Mrs. Burch's health was not good. In December of 1942, Denise and Pamela, sixteen and seventeen respectively, were paddling with some local boys in Middle Harbour, Bantry Bay. Suddenly Denise was grabbed and dragged underwater by a shark. The boys took up sticks, stones, and an oar to try and drive the shark away. The creature let go and Denise surfaced. The girl had suffered horrific injuries and was dead before they could get her to shore.

The attack happened just yards from where thirty-year-old Zita Steadman had been killed by a shark earlier in the year. Zita Steadman arrived by motor launch with a group of friends for a picnic at Egg Rock. They were all in the shallows, but she was a little farther out than the rest when she was attacked by the shark. One of the men in the party, Mr. Bowes, grabbed an oar from the boat and beat at the shark. The shark struck the young woman repeatedly, dragging her into deeper water. Bowes leapt into a rowing boat and tried to ram the shark and was finally able to grab the young woman by the hair and free her body from the shark. She had been bitten in two.

Sharks are elasmobranch fishes. Instead of bone, their skeletons are made of cartilage. The group also includes rays and skates. They range in size from the dwarf lantern shark (*Etmopterus perryi*), at around six and a half inches, to the plankton-feeding whale shark (*Rhincodon typus*) at over sixty feet. Sharks first appear in the fossil record some 439 million years ago in the early Silurian Period. Today, over a thousand species exist, yet only a scant few of these have ever attacked man.

Sharks are marvels of evolution, with astounding adaptations. The ampullae of Lorenzini are electroreceptors possessed by sharks (and some other aquatic animals). These are mucus-filled pores in the skin attached to bulbs containing multiple nerve fibres. They are found mainly on the snouts of sharks and can detect the electrical fields given out by other animals' activity. Even in darkness, a shark can detect potential prey. The ampullae of Lorenzini can also detect magnetic fields and may help sharks and rays migrate.

Sharks also possess a lateral line. This is a system of sensory organs found in fish and used to detect movement, vibration, and pressure gradients via displacement of water. The lateral line is used for navigation and sensing of prey and predators.

Sharks' olfactory organs are packed with plates that are covered with chemical receptors. A shark can detect one part per million of blood in seawater. This is the equivalent of detecting a single drop of blood in an Olympic swimming pool!

THE GREAT WHITE SHARK

The largest of all living predatory fish, the great white shark (*Carcharodon carcharias*), is the most infamous and iconic of shark species. It's the antagonist of Peter Benchley's 1974 novel *Jaws* and, two years later, Stephen Spielberg's phenomenally successful film adaptation (which was one of the few cases of the film being better than the book).

Long before it held us spellbound on the silver screen, the great white swam through the seas of ancient legend.

In Māori mythology, Kawariki was a princess who fell in love with a peasant boy, Tutira. Her father, a sorcerer king, hated the thought of his daughter marrying a commoner, so he cursed Tutira, turning him into a shark. Rather than be defeated, the two still met in secret and would swim together at night. One day, there was a huge tsunami that destroyed the village and swept all the villagers out to sea. Tutira, as a shark, saved the villagers and brought them back to shore. Once Kawariki's father realized the shark that had saved them was Tutira, he was so impressed with this heroic act, he turned Tutira back into a human and apologized by letting him marry Kawariki.

In Greek mythology, the Libyan queen Lamia was the daughter of the sea god Poseidon. She was sexually pursued by Zeus. Hera, jealous of the mortal woman, stole her children, causing her to go mad. Zeus, in a seemingly senseless act, then transformed her into a monster shark that would devour the children of others. Lamia later birthed the many-headed sea monster Scylla. The name Lamia is where we derive the name lamnid sharks from. They are the group that contains the great white.

Another Greek shark demon was Akheilos. He was once a handsome youth, who boasted that he was more beautiful than Aphrodite, the goddess of beauty. As punishment, she transformed him into a hideous shark monster.

The great white is found in all seas and oceans where the water temperature falls between 12 and 24 degrees Celsius (54 and 75 degrees Fahrenheit), from temperate to tropical regions. It feeds on a wide variety of prey, including fish like tuna and rays. Adults prey mainly on marine mammals such as dolphins, seals, and sea lions, favouring items with a high fat content.

The current size record for the species is around twenty to twenty-one feet and is held by a female named "Deep Blue" discovered off Guadalupe Island, Mexico. Another huge specimen was a twenty-foot animal caught off Ledge Point in Western Australia in 1987. Another twenty-footer was caught the following year by David McKendrick off Prince Edward Island and confirmed by the Canadian Shark Research Centre. But, as with most animals covered in this book, these are unlikely to be the largest of their kind.

On April 1, 1987, a twenty-three-foot great white was captured off Kangaroo Island, Australia. Fisherman Peter Riseley and his crew accidentally caught the creature in gill nets. It was too heavy to be brought aboard, so they measured it against their boat.

Amazingly, on April 17 of that year, another huge great white, measured at twenty-three feet four inches, was caught off Malta by Alfredo Cutajar. The fisherman had set lines for swordfish and tuna, and was amazed to find a monster shark entangled in the lines. The beast was longer than Cutajar's eighteen-foot boat. He dragged the monster fish back to the harbour at the village of Wied-Iz-Zurrieq. A larger boat was needed to drag it, dead, around the headland to the larger port of Marsaxlokk.

In January of 2016, the crew of a shark-spotting helicopter released photos of a twenty-three-foot great white they had sighted around a hundred metres off the coast of Marino Rocks, south of Adelaide, Australia. One crew member said it was the biggest white shark he had ever seen. They used a twenty-foot boat as a comparator to gauge its size. The beach was swiftly evacuated, but the shark swam away into deeper water.

During the 1970s, '80s, and early '90s, a huge great white, reckoned to be over twenty-three feet, haunted False Bay in South Africa. It was nicknamed "Submarine" by locals. Shark expert Craig Anthony Ferreira saw Submarine four times as a boy with his father. The shark escaped each time an attempt to catch it was made.

The first encounter saw Ferreira Sr. hook the shark after baiting it with redfish, but after a prolonged battle, it escaped, straightening the hook in the process.

In the second encounter, the hooked shark became entangled in the line of their boat. As it attempted to escape, it damaged the hull and even the intervention of another fishing boat was not enough to subdue it, causing Craig's father to cut the line for fear of being killed by the struggling animal. After that, Submarine escaped and was never hooked again. Despite this, it was seen twice more by Craig and his father, including once when the shark breached several times in succession.

In 1982, ichthyologist Dr. Juan Antonio Moreno saw a gigantic great white brought ashore in Dakar, Senegal. He had no camera or measuring equipment with him, so he measured it using his feet, twice over, and concluded that it was over twenty-six feet long!

This was not the first time such a giant was reported. Ichthyologist Hakan Kabasakal, in a paper on records of great whites off Turkey, records a monster shark captured in the Bosporus Strait in 1926. It was twenty-six feet long and had in its belly two tuna and a dolphin.

Kabasakal also records another twenty-six-foot great white in the Bosporus, caught off the Ahirkapi coast by two fishermen in December 1958, after it attacked and damaged their boat.

The Bosporus seems to be a hotspot for very large great whites, as Kabasakal reports a number of twenty-three-footers caught in the area. Again in 1958, one was caught off Ahirkapi but escaped from the hook and attacked a fishing boat. In the same year, another (or the same) twenty-three-footer was caught off Prince Islands by fishermen Niyazi Dalgin, Cemil Unalir, and Sadan Salvarli, then landed at Ahirkapi.

During an interview Kabasakal conducted with Irfan Yurur—one of the surviving tuna hand-liners who were active between the 1930s and 1990s in the Sea of Marmara—he stated that, due to the discarding of hundreds of tons of surplus bluefin tuna and bonito in coastal waters in the winter of 1958, large sharks, especially great whites, appeared in the Bosporus Strait for an unusually long time, much longer than in previous years.

Due to the collapse of the tuna population in the Bosporus and the Sea of Marmara, great whites vanished from the area. The last one was caught in 1985. They still inhabit the Aegean Sea.

On November 22, 1932, a large white shark was found trapped in a herring weir at the Harbour de Loutre on Campobello Island, Canada. It was said to be twenty-six feet long and six feet in diameter.

Another monster shark supposedly trapped in a herring weir was discovered dead in June of 1930 off White Head Island, New Brunswick, Canada. The fish was measured at thirty-seven feet! This report was given to Dr. Vadin Vladykov, a Ukrainian ichthyologist, second-hand by a fisherman. The creature may have been a misidentified basking shark (*Cetorhinus maximus*), a harmless plankton-feeding shark that can grow to over forty feet long.

However, Canadian Don E. McAllister was able to contact Vladykov on July 11, 1984, when he confirmed that the specimen was a white shark and that the fishermen had given Vladykov two teeth said to be from the shark. One was marked "1936," suggesting it was not from the White

Head Island monster. Its owner was estimated to be about seventeen feet long. The second tooth was larger and undated.

The liver of the giant shark was recorded to have yielded 210 gallons of oil. Using this to estimate its size, a length of twenty-seven feet has been extrapolated.

The great white has been involved in more attacks on humans than any other shark. They seem to fall into two categories: predatory attacks, where the victim is eaten, and exploratory bites, where the human prey is spat out again after the initial attack.

Herodotus, in 492 BC, gives the first written account of great whites attacking people. He tells that the Persian fleet was wrecked at the headlands of Athos in northeastern Greece. As sailors floundered, they were eaten by sharks. During the same period, a Greek tragic poem tells of how a sponge diver, whilst climbing back aboard his boat, had his leg bitten off by a huge shark. The poet notes that the man was buried both on land and at sea!

One of the earliest English-language references to shark attacks occurs in a 1580 *Fugger News-Letter* (so called because they were collected by two brothers named Fugger) which gives this eyewitness account of a seaman virtually falling into the jaws of a shark, somewhere between Portugal and India:

When a man fell from our ship into the sea during a strong wind, so that we could not wait for him or come to his rescue in any other fashion, we threw out to him on a rope a wooden block, especially prepared for that purpose, and this he finally managed to grasp and thought he could save himself thereby. But when our crew drew this block with the man toward the ship and had him within half the carrying distance of a musket shot, there appeared from below the surface of the sea a large

monster called Tiburon, it rushed on the man and tore him to pieces before our very eyes. That surely was a grievous death.

Brook Watson was an Englishman born in Plymouth who was orphaned at the age of six. He went to live in Boston, Massachusetts, with his uncle, who was a merchant trading in the West Indies. Brook became a crew member of one of his uncle's ships.

Whilst swimming in Havana harbour, Cuba in 1749, aged fourteen, he was attacked by a shark that bit off his right leg. The boy not only survived, but became a successful merchant, returned to England and became Lord Mayor of London.

The attack was immortalized in a painting by the artist John Singleton Copley, entitled *Watson and the Shark*, in 1778. The shark is clearly a stylized great white.

When the British maintained prison colonies on Tasmania in the early nineteenth century, prisoners would sometimes escape via a narrow peninsula. The slipped into the sea, swam past the patrolled area, then waded ashore and crept through the undergrowth to freedom. The governor of the colony ordered that garbage be dumped every day in the waters along the peninsula. Lured by the daily promise of free meals, sharks began congregating in the waters of the escape route. After a few screams in the night, and after the prisoners learned about their hungry new watchers, the escape attempts stopped.

It was not till the twentieth century that shark attacks became more widely known beyond the maritime professions. The great white was involved in many of these.

On July 23, 1926, twenty-year-old Augusto Casellato was swimming with his friend Luigi Baldi in the Golfo di Genova of the Ligurian Sea at Regina Elena at San Nazzaro, a resort in Varazze, Italy.

The pair were about half a mile out when Baldi heard his friend cry out. He saw Casellato pulled beneath the water by a huge shark. A boat full of tourists saw the event and rescued Baldi. The shark was said to be twenty-three feet long; its tail fin alone stood five feet out of the water. Casellato's bathing suit and cap were found the next day.

Sometimes great whites will attack in surprisingly shallow water. Frank Athol Riley, seventeen, was swimming off Dee Why, New South Wales, Australia, on March 12, 1934. He was only sixty feet from the shore and the water was only three feet deep. There were about fifty people in the water that day. Among them was Frank's friend Laurie Shields.

> ***Frank was about four feet from me. Then I saw a fin and part of the back of the shark between Frank and me. The shark circled Riley. The shark was so close that I could have put my hand on its back; I had hardly realized what had happened when the shark attacked. Frank had no time to call out before he was dragged under the water and the water above him turned red. Then he appeared on the surface. The shark still had him by the leg. Frank was splashing with his arms and making desperate efforts to get away to the beach. The shark kept jerking him backwards and forwards.***

Lifesaver Laurie O'Toole ran to help, and the shark let go of Frank's leg. With two other men, he dragged the victim to shore. His legs and buttocks had been bitten off, and he died shortly afterwards. The shark was fourteen feet long and almost certainly a great white.

Lifesavers hunted the shark in a surf boat and hooked it, but it broke away. It continued to cruise along the beach for several hours and was seen by beachgoers. Dee Why Beach remained closed for several days afterwards.

Laurie O'Toole was later awarded the Royal Humane Society of Australasia's gold medal and the Surf Life Saving Association's silver medal, its highest award.

It was a warm, hazy afternoon on July 20, 1956, when Tony Grech, an eighteen-year-old Maltese man, was strolling the beach at St. Thomas Bay, just south of Malta's capital, Valletta. He saw his old English teacher, Jack Smedley. Jack had been in the navy and posted to Malta in WWII. After the war, he and his wife Gladys elected to stay on the sunny island, and Jack got a job teaching English. Tony had always liked Jack, who was good-humoured and kind with his pupils.

Fond of swimming, Jack invited Tony to join him in the sea and Tony agreed. The pair swam out into the bay, chatting happily. Suddenly, Tony heard Jack cry, "Look out!" Turning round, he could only see water. Then he felt something bump his chest. Looking down, he saw a huge marine animal. He touched it as it swam past him, and later said it felt like a wet horse. As the dark, shadowy form passed by, Jack surfaced again on the other side of Tony, screaming for help, before the water turned red.

Tony swam madly for shore, where witnesses at the bay and on the cliffs had seen a massive tail and fin. A motorboat took Tony back out to the spot where the attack had happened, but nothing was found. Two days later the search was called off.

A fisherman came forward and said that a few days earlier, he had seen a massive shark swimming past his boat, heading in the direction of Il Ponta tal Munxar.

Tony described the attacker as a huge fish with dark upper parts and a greyish white belly, clearly a great white shark. A wave of panic struck Malta as beaches were evacuated, newspapers ran stories of the danger, and priests warmed against swimming from their pulpits.

A strange conspiracy theory spread that Jack Smedley had been assassinated by Russian agents in a Cold War conflict. Some said he was an undercover agent for MI-5, and a mini sub or divers poisoned or abducted him from below the waves. But clearly Jack was taken by a very big great white. Remember, it was off Malta that a twenty-three-footer was captured in 1987.

Jack Smedley's wife Gladys left Malta a broken woman and died soon after.

Great whites will enter tidal river mouths in search of prey. On December 28, 1927, in the Little Brak River, Western Cape Province, South Africa, seventeen-year-old Ockert Stephanus Heyns was bathing with a group of friends when a big shark surfaced in the middle of the group. As the others made it to shore, the shark flew at Heyns like an arrow, biting his left leg clean off above the knee. Heyns's uncle, Muller Heyns, and spectators dashed into the reddening water and the shark swiftly made off. The attack took less than fifteen seconds.

He was taken to Mossel Bay Hospital, but died forty minutes after the attack. The *Mossel Bay Advertiser* of December 31, 1927, reported that two days after Heyns's death, a shark was caught on a pork-baited line and relics of the victim were found in it. It was sixteen feet four inches long.

Mr. A. Muller Heyns was presented with three decorations for bravery: He was awarded the silver medal of the Royal Humane Society, the gold Stanhope Medal for the bravest deed of the year, and the bronze Albert Medal.

On November 1, 1942, eighteen-year-old medical student Willem Johannes Bergh was swimming off Clifton, Western Cape Province, South Africa. He had several friends with him. Berg had only swum a few strokes when a titanic great white shark attacked him. He struggled free from the jaws momentarily but was snatched again. The monster shark swam back out to sea with its victim. One witness was Gary Haselau, who later wrote of the event in the 1971 book *Underwater Africa*, edited by Al J. Venter. Haselau estimated the shark to have been twenty-three feet long.

Twenty-five-year-old Peter Savino had been swimming with Daniel Hogan, twenty-two, off Atascadero Beach, Morro Bay, California, on April 28, 1957. Both men had been swept some 1,800 feet from shore by a strong ebb tide. Savino had become exhausted from fighting the current and was being towed back to shore by his friend. Hogan told Deputy Sheriffs Don C. Miller and Henry Karagard what happened next.

Pete had gotten tired and was hanging onto my shoulder when there was a sudden swirl of water and he disappeared over the

top of a big wave. I heard him yell, "Something really big hit me," and I saw him raise an arm out of the water that was all bloody. I saw the shark hit him. I said, "Come on, let's get out of here," because I knew the blood would bring the shark back. I saw the shark, it boiled the water around us and then it all got confused, but I saw the shark. It carried Pete over a big wave. I started swimming toward shore and Pete was right behind me. The next time I glanced back he was gone.

Hogan swam to shore and got a friend, Jerald Frak, to phone the police and Coast Guard. The Coast Guard dispatched the cutter *Alert*, which, upon arriving on the scene, lowered a twenty-one-foot launch under the command of executive officer James C. Knight. Within mere minutes of beginning their search for the missing swimmer, Knight reported, "We located a shark as long as our launch. After making a quick trip back to the *Alert* for firearms we returned to the area where we had last seen the shark, but it was gone."

The shark was never captured, and Savino's body was presumed utterly consumed by the fish.

Robert Lyell Pamperin, thirty-three, was free-diving with his friend Gerald Lehrer, thirty, off La Jolla Cove in La Jolla, San Diego County, California, on June 14, 1959. The men were collecting abalone (*Haliotis rufescens*), a species of edible sea snail whose shell is also used in ornamentation. The men had got separated when Lehrer heard his friend cry for help. Turning round, he was baffled to see Pamperin seeming to stand up unnaturally high in the water. Then he saw the water turn red and his friend vanish below the surface. Diving under, he saw Pamperin in the jaws of a twenty-three-foot-long great white shark. He told the *San Diego Union* newspaper, "It was so big I thought at first it was a killer whale. It had a white belly and I could see its jaws and jagged teeth."

At the same time, William Abitz stood on an elevated rock formation that overlooked the site. He was alerted to the event by Pamperin's cries for help. Abitz recounted, "Pamperin was thrashing as if he were trying to run away from something, then he disappeared below the surface."

Lehrer bravely dived under again to try and rescue his friend, who was being swallowed feet first. Realizing nothing could be done, Lehrer swam towards the beach and was met fifty feet from shore by William Abitz, who had swum out to help. Upon reaching shore, they located lifeguards, who informed local authorities.

Divers scoured the seabed, and a helicopter searched the surface, but no remains were found. The shark had seemingly swallowed its victim whole. It later emerged that fewer than two hours before the attack, several yellowtail (*Seriola dorsalis*) were speared by divers near the attack site. Also, a US Navy sailor swimming off nearby rocks had badly lacerated himself, losing a considerable amount of blood while in the water. A dead whale had washed up less than half a mile from where the men were diving. Add to this that there was a colony of harbour seals (*Phoca vitulina*) nearby, and you have the perfect storm for attracting big sharks.

Sometimes irony rears its head. Maurizio Sarra was a twenty-eight-year-old underwater photographer and author of the book *My Friend the Shark*. On September 2, 1962, he was spearfishing in the Tyrrhenian Sea at Secca del Quadro south of Monte Circeo, Italy. He was scheduled to leave on September 10 for Polynesia on an assignment for the University of Rome.

Sarra had speared many fish, which he had attached to his belt. A motorboat approached the boat from which Sarra was diving to advise them that a white shark had been sighted in the area. He unloaded his fish, then dived down again to retrieve the speargun and camera that he had left on the seabed. As he resurfaced, the shark bit him. Apparently, he did not know how badly he had been mauled, joking, "These sharks bite hard." However, once he had been hauled onto the boat, he lost consciousness. Both of his legs had been badly bitten, bone deep from thigh to heel.

It took forty minutes to get him to the hospital. Despite being given blood and four hours of surgery, Sarra died at 11:30 that night. With friends like that...

A dramatic case, worthy of an action movie, unfolded on August 19, 1967, off Jurien Bay, Western Australia. Robert Bartle and Lee Warner were taking part in a spearfishing competition with other divers. Lee remained on the surface whilst Robert dived.

The ocean floor was barren, so Robert looked for some limestone caves where there might be fish. Finding none, he ascended again and was struck by a big great white. It bit him between the hip and shoulder and shook him violently. Lee shot a spear from his gun into the shark's head.

The shark bit Robert clean in two, then rose up towards Lee.

The shark kept circling around me with the body of Bob still in its jaws. I could see the terrible wounds which had been inflicted. I felt helpless. I could see Bob was dead... It kept circling about three metres from me... Out of the corner of my eye I saw Bob's gun, which was still loaded and floating just below the surface. I grabbed it thinking, here was another chance! Swinging it around, I tried to belt the spear into the shark's eye.

He shot the spear at the shark, but it missed. The fish kept circling. The shark continued to circle until it became entangled in the float and spear lines, and a small bronze whaler shark (*Carcharhinus brachyurus*) began darting around. Warner, defenceless with no spearguns, began swimming backwards and finally reached the shore.

Unable to find the keys to Robert's car, Lee found keys in a car belonging to one of the spearfishers who was still in the water. He raced to the small fishing settlement at Sandy Cape. There he found Harry Holmes, skipper of the forty-four-foot *Gay Jan*. The small boat sped to the headland and plucked the divers from the water. They also found the shark, still tangled in the floats, and Bartle's torso. As they tried to pull the shark towards the boat, the lines snapped, and it swam away.

Not all great whites that attack humans are huge. Such was the tragic case of twelve-year-old Wade Shipard. The boy was swimming at Point Sinclair, near Cactus Beach, Australia, behind a sheltered headland where shark fisherman George Mastrosavas and his fishing partner were gutting and cleaning school sharks (*Galeorhinus galeus*) they'd caught at sea. It was February of 1975.

Mastrosavas suggested to his mate that they shouldn't gut the fish there, as kids swam in the area and the blood might attract sharks. His mate, the skipper of their fishing boat, replied, "Blow the kids! They can swim in the dam."

Shipard was swimming in the vicinity of crayfish pots containing material that had been proved to attract sharks, and where fishing boats routinely unloaded their catch and washed down their boats, increasing the amount of blood and offal in the water.

Mastrosavas said, "We saw this young lad swimming, and next thing he was yelling that a shark had bitten him. We sped towards him in the boat, but before we got there we saw the shark—a small white—come up behind him and swallowed his leg. It just rolled and snapped his leg right off."

The shark was only ten feet long. Mastrosavas slashed the shark with a gaff as it came through the boy's blood for another strike, then pulled the boy into the boat.

However, he was unable to staunch the bleeding.

I ran down the jetty with him in my arms, and his mother asked me to drive their car because of the state she was in. We drove like mad towards Ceduna, and after a while the lad sorta sat up. He called out, "Mum!" And that was the end—he died in the car.

Mastrosavas blamed himself for the events and later had a mental breakdown, spending ten years in an institution.

A shark of gigantic size was implicated in the death of Lewis Archer Boren, twenty-four, on December 19, 1981. Boren and a friend had been surfing at South Moss Beach, Spanish Bay, Monterey, California. At two in the afternoon, they decided to call it a day, and his friend waved goodbye and left. That was the last time anybody saw Boren alive. Apparently, Boren decided to catch one last wave alone. It was a decision that led to his death in the jaws of a truly huge shark.

Two surfers found Boren's surfboard washed up the following day with a massive bite mark in it. Analysis showed teeth marks from an immense great white. Then, at eleven o'clock in the morning on December 24, Boren's body was spotted by a park ranger near Pacific Grove. The body had a bite mark extending from beneath his left armpit all the way to his hip and extending halfway across his body. Boren was around six feet tall. Examination of the bite radius by experts led them to conclude that the shark that killed Boren was in excess of twenty-three feet long.

A fishing company had been dumping offal into the sea at Wiseman's Beach in Peake Bay, north of Port Lincoln, Australia. The offal attracted a massive great white shark that had been seen in area for about a month from early February of 1985.

The great white is not afraid to attack boats, as Tamara McAllister and Jeffrey Stoddard found out on January 26, 1989. The pair, both twenty-four, were kayaking off Paradise Cove, west of Malibu, Los Angeles County, California. At around nine in the morning, they were seen eating muffins and drinking coffee before launching their kayaks and paddling around Latigo Point as they headed north towards Paradise Cove at nine thirty.

At about 10:15 a.m., Margaret Bloom, a resident of Paradise Cove, returned home from an early morning doctor's appointment and was standing in front of her living room picture windows, looking out at the cove. She observed the water seeming to boil and thrash beyond the kelp beds, close to a Coast Guard buoy. Sea lions were jumping up onto the buoy. She said, "There was a lot of splashing water and a churning of the ocean. It was like a whirlpool, maybe 15 to 20 feet across. It lasted about five to ten minutes, then stopped, with all going quiet in the water."

Bloom said the pinnipeds on the buoy appeared to be trying to crawl up on top of it. They were very agitated when the water was being churned up.

Next day, the couple's kayaks were found, lashed together, off Zuma Beach, almost four miles away. They were turned over to the Ventura County Sheriff's Department. One was found to be fractured underneath and at the sides. It was worked out that something had slammed into it from underneath. That something would have been travelling at seventeen knots and massed at around a ton.

On January 28, Tamara McAllister's body was found. She had been killed by a massive bite on her upper left thigh. Stoddard's body was never found.

Investigators believed that the pair had tied the kayaks together due to the sea becoming choppy. A great white shark, calculated to be sixteen feet long, had attacked Stoddard's kayak from below, tossing him into the air and upsetting McAllister's kayak. The shark bit McAllister fatally and may have eaten Stoddard.

Forty-one-year-old professional diver Kazuta Harada was fishing for pen shells (*Atrina ectinate*) off Horie, Matsuyama City, in Ehime prefecture, Japan, on March 8, 1992. He was diving at a depth of around 126 feet in the Seto Inland Sea. A rescue rope, a rubber-coated radio cable, and an air tube connected Harada to the support boat.

At 3:20 in the afternoon, he shouted via the radio to be pulled up. Crew on the boat were unable to pull the diver up by the rope. They also tried unsuccessfully to pull the air tube. Then they tied the air tube to the boat and pulled it by moving the boat very slowly. In this way they hauled in the air tube, but the rescue rope and the radio cable had been severed by then. Eventually the air tube was pulled in. At the end of it was an empty and severely damaged diving suit and helmet.

An extensive search failed to find Harada. Marine biologist Kazuhiro Nakaya examined the diving suit. The right leg and right trunk were missing, and the left leg turned inside out, suggesting the man had been violently ripped from the suit. A fragment of shark tooth was taken from a metal shoulder protector. The killer was identified as a great white shark sixteen feet long.

Other divers had reported a sixteen-foot shark in the same area in the two years before the attack. Some of them had been bitten but were protected by their steel helmets. Two boats had been approached and one attacked. Yoshiaki Ueda, a mackerel fishermen, had his boat rammed by a great white. The shark bit at the boat as the skipper tried to fend it off with a wooden pole for five minutes. The shark left two teeth from its lower jaw in the boat.

Debbie and John Ford had been married for only fifteen days. They were on their honeymoon in Byron Bay, New South Wales, Australia. It was June 9, 1993. The couple had joined a group of people diving off Julian Rocks within the Cape Byron Marine Park.

Whilst underwater, John saw an eighteen-foot great white swimming directly towards his wife. Courageously pushing her out of the way, he placed himself in front of the beast's jaws. The shark bit down on his torso.

Professional diver Sally Gregory was on a nearby boat when the skipper of Debbie and John's craft called for help. Debbie burst to the surface in hysterics. The shark was still circling. Sally risked her own life to dive into the bloodied water and swim to Debbie. She comforted the distraught woman on the boat as Sally's best friend and fellow diver Geoff Brackenrig dived into the depths to see if John could be saved.

Sally recalled, "One of the divers was a concert violinist and we had snuck his instrument on board, so as people were returning to the surface he was playing Beethoven's Ode To Joy. It sadly turned into a requiem for John Ford who died protecting his wife from a shark."

Sally and Debbie were told to stay on the boat and not move. She found out why later. "I learned later that his remains were floating to the surface and the men in the other boat were collecting them in a bucket. They didn't want us to see it."

Local fisherman Ron Boggis was later enlisted to hunt the shark, which he described as a monster. It was hooked with a bait line 3.7 miles out at sea, but it bit through the steel fishing cable. It vomited up John's body after an hour and a half struggle. Some days later it was seen rolling in the waves at Hastings Point and is believed to have died of the wounds it sustained.

Some divers use anti-shark equipment to deter attacks. Shark defence pods are devices that consist of two electrodes that emit a three-dimensional electrical field surrounding the user or area. When a shark comes within a few metres, the field emitted by the device causes it to experience muscle spasms due to interference with the ampullae of Lorenzin.

When Paul William Buckland and Shannon Luke Jenzen went scallop diving off Saddle Point, Australia, they took a shark defence pod with them. It was April 30, 2002.

They had a hookah unit, a battery-powered air compressor that delivers oxygen to the diver via a hose. They also had a shark defence pod and a twenty-foot boat.

Jenzen dived first, for about an hour, with the pod switched on. He then handed the device to Buckland as they swapped over. Jenzen noticed that his friend had not switched the pod on before entering the water. Buckland had only been underwater for five minutes when he surfaced in distress. Jenzen said:

> ***I was sorting a catch and I was about three quarters of the way down the hose when I heard Bucky calling out to me. He was yelling my name. I didn't see him come up, but I knew something was wrong. I think he was saying, "Shannon, come quickly..." I kicked the motors over and put it in gear and started to motor over to him.***

> *The shark attacked virtually straightaway and I saw it happening in front of me. I saw that it was a white pointer and it had Bucky in its jaws. It lunged out of the water and was shaking its head and thrashing around. It didn't take Bucky under. I steamed up to the shark and it was still attacking him; it wouldn't let him go. The shark was enormous; it was the size of the boat. The girth of it was huge. It was just thrashing him on the top of the water. Bucky wasn't screaming at all, it was just so violent. I reached him and hit the shark with the side of the boat. I went around to the steps at the side, got on the steps, and pulled Bucky in. The shark let go as I pulled Bucky in. As I pulled Bucky from the water, I felt the SharkPOD zapping me. If the POD is on it will give out little electric shocks when it comes out of the water. Bucky was still alive when I was pulling him in, but I saw that his injuries were extensive. As I pulled him in, he just said, "Get me in the boat." He died pretty well straight after I had got him in.*

The radio on the boat was defective, and the mobile phone had a flat battery, so Jenzen was unable to contact the shore. He steered the boat back towards Smokey Bay, stopping on the way to ask some fishermen to radio for an ambulance, which they did. He warned other fishermen in the area to get their divers out of the water. He then headed to the boat ramp and awaited the ambulance, but it was clearly too late to save his friend.

Buckland's right leg had been totally severed and there were extensive bites to his buttocks and upper left leg.

Apparently, they only turned the pod on when ascending or descending. The shark seemed to have attacked when the pod was switched off.

A huge great white had been harassing boats in the area during the two months prior to the attack on Buckland. It was said to be twenty to twenty-three feet long. It may have been the individual involved in the attack.

Seventy-seven-year-old Tyna Webb had swum off the same beach for seventeen years, False Bay off Fish Hoek Beach, Western Cape Province, South Africa. She swam six days out of seven. On November 15, 2004, she was enjoying her favourite pastime when she backstroked into the path of a massive great white. A dozen witnesses saw the events, including fisherman Jeffrey Andries, who saw the attack through binoculars. Another was Tim Atkins, in a car high on the mountainside. He said:

> *I suddenly saw a shark coming at great speed from the Kalk Bay end of the beach towards where the woman was swimming. The shark hardly slowed down. It just hit her and the water was full of blood. It made two turns, grabbed her in her side and pulled her under the water. The shark then turned and headed out to sea. I think it had the woman in its jaws.*

False Bay Yacht Club rear commodore Paul Dennett called the rescue services after seeing the shark thrashing the woman violently in its jaws.

> *It was thrashing her body in the water and then swam off with her. It came around and its whole mouth came out of the water, and not even breaching the water took her down. All that was left was a little red bathing cap about 100 metres from where the bloodied water was dissipating.*

Two National Sea Rescue boats were launched as well as a helicopter. NSRI spokesperson Craig Lambinon reported that the helicopter spotted a large shark in the vicinity of Clovelly, near Fish Hoek. "The shark is bigger than the helicopter...it is huge," he said. The shark was estimated to be twenty feet long. The victim's body was never found and most likely had been eaten whole by the shark.

Fishing boats returning to the area had formed an accidental chum trail due to the blood from their catches being pumped into the sea from their bilges and washing off their decks. Shark tour operator Theo Ferreira had been chumming close to the beach and later lost his licence.

A year later, in June of 2005, False Bay again became the scene of a fatal shark attack, this time off Miller's Point. Twenty-two-year-old medical student Henri Murray was spearfishing with friend Piet van Niekerk. Henri speared a black bream (*Coracinus capensis*). Van Niekerk was thirty feet away when Henri shouted, "A great white is here; we have to get out!" Van Niekerk later said:

> ***While we were trying to get ashore, the shark jumped out of the water next to us. He disappeared under the water again. At first, he snapped at Henri twice, but was unsuccessful because Henri fought him off. The third time the shark pulled him under the water. I dived beneath the water and I saw Henri's arms in the shark's mouth.***

Van Niekerk shot the shark with his speargun without effect, and the shark swam off with Murray. Van Niekerk swam to shore and ran for help. Some fishermen came to his assistance, used their boat to search for Murray, and called for help. Emergency services responded, including the National Sea Rescue Institute, Metro rescue, and Simon's Town police and fire services.

Dave Estment was sitting on a jetty and saw the attack.

It was incredibly fast. Suddenly a huge shark surged from under the water, taking the one diver up to his arms in its jaws. It must have been massive to have done that. Then the shark and the man just vanished.

Grant Munro, another witness, saw the shark lift Henri clear of the water in its jaws. He estimated it to be sixteen feet long.

Police divers recovered the speargun, a flipper, a mask and snorkel, and part of a weight belt belonging to Murray. The shark was seen next day off Roman Rock Lighthouse in Simon's Town and in Kalk Bay harbour by fishermen. The spear fired by Piet van Niekerk was still embedded in the shark's body, and the shark was towing his speargun.

British grandmother Doreen Collyer, sixty, was scuba diving off the northern Perth suburb of Mindarie between One- and Three-Mile Reef in June 2016. Her dive partner, John, felt something swim past him in the water, then saw a commotion on the surface.

Three men heading out to fish saw the shark and helped John onto their boat, then positioned it between the shark and its victim so they could pull the body aboard. Doreen was already dead. They said the shark was longer than their eighteen-foot boat.

Just two days before, a great white had killed surfer Ben Gerring by biting his leg off at Falcon Beach in the town of Mandurah, south of Perth.

Simon Nellist was a fifty-five-year-old diving instructor from Cornwall in the UK. He was a former Royal Air Force gunner who had survived two tours in Afghanistan. He had moved to Australia and was soon to marry. In February of 2022 he was swimming off Little Bay, New South Wales. He was attacked from beneath by a fifteen-foot great white and dragged out to sea.

A search with helicopters, boats, and jet skis later found some of his remains. It was thought that fishing in the area had attracted the shark.

My own cousin, Paul Bolstridge, had swum in that very same place just weeks before the attack.

THE TIGER SHARK

Widely regarded as the second most dangerous shark, *Galeocerdo cuvier*, or the tiger shark, is a fish of tropical and subtropical waters. Unlike the great white, it has little tolerance of cool water. The fish is named after the striped markings along its sides.

As with most, big, dangerous animals, legends about tiger sharks abound.

Kamohoali'i was a Hawaiian shark god. He would swim in front of boats lost at sea and, when rewarded with a narcotic brew called *awa*, he would guide the ships home.

Ukupanipo is another shark god from Hawaii. He controlled the amount of fish humans could catch. Sometimes Ukupanipo adopted a human child who gained the power to transform into a shark.

In some parts of Hawaii, it was said that sharks would be trapped in rock pools and warriors would prove their worth by fighting them with spears tipped with shark's teeth.

On the Solomon Islands, "shark callers" call up sharks and sacrifice pigs tied to wooden stakes to them. It is beloved that the sharks will not attack them, and children play in waters where sharks swim. The shark caller is said to be able to send spirit sharks to kill anybody who has wronged them. These spirit sharks can take on human form and hunt their victims on land.

The tiger shark can reach eighteen feet in length, though most are smaller. In August 2015, Geoff Brooks posted two pictures on Facebook of a tiger shark caught off Seven-Mile Beach near Lennox Head, New South Wales, Australia, said to be twenty feet long. A fisherman calling himself "Matthew" came forward to confirm that it was he who captured the shark. He had hooked a hammerhead shark when the huge tiger shark attacked and ate it.

I was the one that took that photo and I was the one that caught that fish. I was fighting the hammerhead shark and the tiger shark came up and swallowed it.

He also claimed to have seen twenty-three-foot tiger sharks off the coast. Matthew said that the shark was taken to a fish market and cut up whilst he retained the jaws. Another report said that the body had been sent to the Commonwealth Scientific and Industrial Research Organisation (CSIRO). However, a spokesperson told *Daily Mail Australia* that the shark has not been accounted for by CSIRO staff.

"We've seen the photos, but we haven't got any information that it has been handed in or where it was caught," he said.

There is one unconfirmed report of a specimen twenty-four feet three inches long caught off Indochina in 1957, but details are lacking.

Tiger sharks have unique teeth shaped like serrated, backwards-pointing crescents. The shape is not unlike an old-fashioned can opener, and the shark uses them in the same way to saw into the shells of the marine turtles that are among its favourite prey. Tiger sharks also feed on marine mammals like dolphins, seals, and sea lions, and a wide range of fish, as well as inedible objects such as oil cans, car licence plates, tires, Wellington boots, and a fur coat!

Tiger sharks can and do attack humans from time to time.

A big tiger shark took a man involved in a boating accident in the mouth of the Mississippi River, Louisiana, in 1879. The story was related some twenty-one years later by a Captain McLaughlin in the *Indiana County Gazette*.

I was out looking for ships with my partner, Captain Tom Wilson, and the usual crew, and about 12 miles out of South Pass we sighted a large sailing vessel which proved to be the Zephyr, from Bath, in charge of Captain Switzer. There was a rival pilot boat nearby, and we both made a rush for the ship to get the job of taking her in.

Our party was nearest, and Captain Wilson and two sailors put off in a small boat to go aboard, but in their hurry they made a miscalculation and were struck by the bow and capsized. It all happened in a flash, but Wilson and one of the sailors were lucky enough to get hold of the overturned boat and hang on. The other sailor was thrown some distance away into the water.

He was a big, brawny, six-foot Swede named Gus Ericsson, and when we saw him come up one of the crew tossed him a circular life buoy, which he seized at almost immediately. The buoy was amply sufficient to sustain him, and he put his arms across it and held himself out of the water fully breast high. We had another small boat and started at once to pick up the three men, making for Ericsson first.

When we were less than 100 feet away, I saw a gigantic tiger shark rise and start toward him, and the next instant the poor fellow shot down out of sight, life buoy and all, like a man going through a trap. We were so horrified that we simply sat and stared, and what seemed to be two or three minutes elapsed. It must have risen from a great depth, because it bounded at least four feet into the air and fell back with a splash. Of Ericsson we never saw a trace. He went into that

shark's jaw as surely as two and two make four. We rescued the other men all right.

On July 25, 1983, a sixteen-foot tiger shark killed two people off Lodestone Reef, Great Barrier Reef, north of Townsville, Queensland, Australia.

The forty-six-foot prawn trawler, the *New Venture*, was swamped by waves. The captain, Ray Boundy, twenty-eight, deckhand Dennis Patrick Murphy, twenty-four, and his girlfriend the cook, Linda Ann Horton, twenty-one, were all thrown into the night sea.

They stayed afloat with a surfboard, some foam, and a lifesaving ring, but a large tiger shark began circling them. The skipper, Ray Boundy, was lightly bitten on the knee. A few minutes later, all three were knocked off their float by a wave. When they resurfaced, a shark had hold of Dennis's leg. He called out, "The bastard's got my frigging leg." The shark then tore off his leg. Blood was pouring out of the wound and Dennis realized the blood would attract other sharks.

Bravely, Dennis swam away from the others, hoping to lead the shark away. He asked Ray to take care of Linda. He was again attacked not far from the float. Ray later said, "He was just screaming and I couldn't believe that anyone could have that much guts to get his mates out... It was like watching a human being fed through a mincer. Linda was hysterical."

At four in the morning, the shark was again noticed circling the float. It suddenly attacked Linda and pulled her down. Ray later said, "It flung itself into the air and got the top half of her and turned her upside down. It was just so quick and she squealed and it shook her like a rag dog... She didn't say anything, but was mumbling."

A few hours later, the shark returned, but before it attacked, Ray caught a wave over a reef and managed to escape. Soon after, a rescue plane arrived, and he was saved.

Richard Peter Bisley worked on a pearl farm off Roebuck Bay, western Australia. The farm raised pearls in the shells of *Pinctada maxima*, a species of oyster that produces the most valuable pearls in the world.

It was November 1993 and Bisley was hookah diving, cleaning the shells of weed. Bisley and Stefan John Freney-Mills were conducting the maintenance together when Freney-Mills's air supply suddenly dropped in pressure. He switched to his pony bottle (a small diving cylinder) and surfaced. Bisley's hose was free-flowing and when his line was pulled in, his weight belt was still attached, but the hose to his demand valve mouthpiece had been severed.

An immediate search of the pearl farm and the surrounding area was initiated. Police were contacted and within forty-five minutes, nine dinghies with divers and crew were searching for the missing diver. A Cessna plane joined the search.

Next day Bisley's face mask, demand valve, and pony bottle with its shoulder strap severed were found on the seabed, and the fence lines on which the pearl shells hung were damaged. Farm manager Kau Stainton said that it looked as if a large, heavy marine creature had crashed through the panels.

A tiger shark was caught near the pearl farm six days after Bisley disappeared. Clothing and human remains in its gut were determined, on forensic examination, to be those of the missing diver.

Liya Sibili, a twenty-year-old man, was swimming at Second Beach, Port St. Johns, Eastern Cape Province, South Africa. It was Christmas Day, and the water was full of festive people. Lifeguards noticed that a large tiger shark had swum in amongst them and tried to get everybody out of the water, but were unable to do so fast enough.

Board surfer Avuyile Ndamase was a witness to what happened next.

> ***People were trying to get out the water after the lifeguards spotted something in the water. I saw the tail of something really big grab the boy, the water started boiling and later it***

went red. After all the attacks we have had here over the years, people are crazy to even go in the water.

Ndamase's younger brother had been killed by a shark in the same bay the year before.

All that was found of Sibili were his trousers. Second Beach lodge owner Mike Gatke said under-equipped lifeguards were extremely nervous to go into the water after witnessing the latest attack. "They are very scared. There is no way they will go into the water if someone is attacked."

In July 2002, two women were attacked and killed by a tiger shark off Hurghada, Egypt. An Austrian woman, sixty-eight-year-old Elizabeth Sauer, had an arm and a leg bitten off by a shark but, amazingly, managed to get to shore before she died. The body of an unnamed Romanian woman was found on a reef the same day. Her body had been mutilated in a shark attack. The same animal may have been involved in both attacks. To have two attacks in the same area on the same day is very rare, but the next year, more horrific events were to play out.

In June of 2003, again at Hurghada, Vladimir Popov, twenty-three, and his father Yury, were spending the day by the Red Sea. The father and son were from Russia and were on holiday at the resort. As he was swimming, a ten-foot tiger shark began to circle Vladimir, then attacked him, dragging him under. He cried out "Papa, papa!" as he vanished. The water turned red as the shark fed on the man's body for two hours.

Yury told the local media local media:

What kind of help can you give? This meat grinder happened in 20 seconds, he was just dragged under the water. This is an absolutely ridiculous coincidence, because it is a safe beach. There are ships and yachts around. It's never happened there.

They usually attack on wild beaches. It's just some kind of evil fate.

But we have seen this is quite wrong. Tiger sharks will swim into crowds of people.

The shark was caught in a net and dragged to shore, where it was bludgeoned to death by a crowd. Authorities closed off a forty-six-mile stretch of the coastline. The shark was found to be a pregnant female. The only food in its stomach was the remains of Popov. It is thought that the attack happened because the shark was starving due to massive overfishing in the area.

THE BULL SHARK

Smaller than the great white and the tiger shark, the bull shark (*Carcharhinus leucas*) is considered one of the most dangerous sharks in the world. Found in tropical and subtropical waters, the bull shark is one of the few shark species that can live in fresh water. It enters a number of tropical rivers and has become landlocked in Lake Nicaragua. There are many local names for it, including the Nicaragua shark and the Zambezi shark.

Most sharks must retain salt in its their bodies to survive. If not, their cells will rupture, causing bloating and subsequent death. For this reason, most sharks will not enter fresh water. Even if they do, they return to salt water quickly; if not, their internal salt levels would become diluted. When they are placed in fresh water for extended periods of time, they absorb too much water and lose too much salt concentration to stay alive.

Bull sharks are thought to enter fresh water to give birth to their pups and thus avoid competition from other sharks. Bull sharks in freshwater environments decrease the salt-excretory activity of the rectal gland, thus conserving sodium and chloride. The kidneys also play an important role in the active reabsorption of solutes into the blood and they also produce large amounts of dilute urine. Bull sharks in fresh water urinate twenty times more than those in salt water. The gills extract salt from the water and the kidneys recycle the salt content within their bodies.

Perhaps, due to its smaller size, there is less folklore about the bull shark than there is about the great white and tiger sharks. Warrau Indians of South America tell a story of how a hunter named Nohi-Abassi trained a shark to attack and kill his mother-in-law. But her daughter found out about the plot and disguised herself as the shark. Instead of attacking the mother-in-law, she cut off her husband's leg. The leg was transformed into the constellation known as Orion's belt.

The "Festival of the Sharks" used to be held annually in some parts of the New Hebrides. The ceremony lasted for a week. The body of a shark was placed in a sort of native altar and was buried. A native artist, with white pigment, painted the figure of a shark upon the grave, which for a time was constantly guarded.

A Catholic priest, the Reverend A. J. Laplante, witnessed shark charming in the Fiji Islands during the decade he spent as a missionary there between 1928 and 1938. Father Laplante said the islanders subdued sharks by kissing them. "It's some occult power they have which I can't define. But once the native kisses it, that shark never moves again."

Twice a year, when the natives made a drive for food for tribal feasts, or when they wanted to make their swimming areas safe from sharks, shark-kissing ceremonies would be held.

> ***The night before the drive, the man who wants the shark fishing done goes to the house of the chief, who is also the sorcerer or medicine man. There they enact a ceremony which survives from their oldest superstitions and beliefs. This ceremony always includes the presentation of kava, a mildly narcotic beverage made from juice extracted from finely ground roots, and the sacrifice of an animal. The kava is drunk and some of***

it is sprinkled on the important main post of the house, where the spirit lives, and the animal is strangled, cooked, and eaten. The next day, the natives drive the sharks into a large net, and the shark-kissers wade out, seize the man-eaters, kiss them on their upturned bellies and fling them on the bank. I don't know how they do it, but, among the natives, it is taken for granted that once a shark is kissed upside down, that is the end of it.

Clearly what is happening here is not magic, but tonic immobility. This is a trance-like state sharks go into when turned upside down. Nobody quite knows how this works or why it happens, but the shark's breathing slows and its fins stiffen. It may remain in this odd biological trance for up to fifteen minutes. Killer whales flip great white sharks and other species onto their backs to kill them more easily.

Bull sharks average around eight feet long, but one outsized specimen caught on the Breede River in South Africa was thirteen feet one inch. It was caught, measured, and released by biologist and angler Jeremy Wade in 2010.

Opportunistic feeders, their diet consists of fish, turtles, sea birds, and mammals. They have been involved in a number of attacks on humans.

On August 9, 1878, sixteen-year-old Arthur Cole was swimming in the East River, New York City, New York. A friend who was in a boat nearby saw him struggling with an eight-foot bull shark and picked up a rock that was used as an anchor and hurled it at the fish. The rock struck the shark on the head, stunning it. The boy then dragged his mutilated friend into the boat and rowed ashore, taking the almost lifeless body to the police. The wounds to his sides and back were so bad that he did not survive.

Ironmonger Joseph Dobson, twenty-nine, was swimming twelve miles inland near Como in the George River, New South Wales, Australia on January 28, 1903. He was swimming alone, but his brother and a couple of friends were on the bank. Dobson was attacked and savaged in the water, and the onlookers saw the shark launch itself out of the water at its victim several times as he cried for help. As he tried to fight the shark off, it took his hand.

The others managed to drag Dobson ashore. His right forearm was missing, and there was a large deep wound on the inside of the right elbow. The abdomen showed teeth marks and the soft parts were torn off from the ribs to the head on the right side. His entrails were exposed. The left leg was broken at the ankle where there was a large wound. The right foot was also badly bitten.

A large bull shark was the suspected attacker.

One of the strangest shark attacks happened in Florida in 1907. Belton Larkin was the owner of a sponge fishing concession. On Monday, March 26, Larkin and another man were in a boat off Garden Key, Pine Island Sound. What happened next was recorded in the *Punta Gorda Herald* of March 31, 1907.

Two fishermen busy with a net did not notice the approach of a tarpon pursued by a shark. In one of its tremendous leaps, the tarpon fell across the skiff. The skiff broke in two and the fishermen became entangled in their net, and the shark took a huge bite out of the side of one of them, Belton Larkin, cutting his body nearly in two.

For twelve days, between July 1 and July 12, 1916, New Jersey became the scene of the most infamous series of shark attacks in history. Such was the impact of this case that, years later, it inspired Peter Benchley to write *Jaws*. The case has been argued about ever since, and it may be unique in the it possibly involved more than one species.

What has since been described as "twelve days of terror" began on Sunday, July 1, at Beach Haven, a resort town in Long Beach Island off the southern coast of New Jersey. Philadelphia resident Charles Epting Vansant, twenty-eight, went for a swim with his dog. When Vansant began to shout, other bathers thought he was calling his dog. In fact, a shark was savaging his leg.

He was rescued by lifeguard Alexander Ott and bystander Sheridan Taylor, who claimed the shark followed the blood trail to shore as they pulled the bleeding Vansant from the water. His left thigh was stripped of flesh, and he died on the manager's desk at the Engleside Hotel.

Amazingly, the beaches stayed open, and warnings from sea captains of observations of large sharks in the area were utterly ignored. But six days later, another resort town, Spring Lake, forty-five miles to the north, became the backdrop of the next attack.

Charles Bruder, twenty-seven, was a Swiss supervisor of bellboys at the Essex & Sussex Hotel. Whilst swimming a shark attacked him, biting his abdomen and severing his legs. Apparently, there was so much blood in the water that a woman who alerted lifeguards thought a red canoe had capsized. Lifeguards Chris Anderson and George White rowed to Bruder in a lifeboat. Apparently, women screamed and fainted as the victim's body was brought ashore. Workers at the hotel raised money for Bruder's mother back in Switzerland.

The next attacks took place in Matawan Creek, which has given its name to the case, a tidal inlet of Raritan Bay, thirty miles north of the last attack. Thomas Cottrell, a sea captain and Matawan resident, spotted an eight-foot-long shark swimming in the creek. When he told people, he was laughed at.

On July 12, eleven-year-old Lester Stillwell and his friends went swimming in an area of the creek called Wyckoff Dock. The boys saw what they thought was a log or dark plank in the water. When a dorsal fin broke the surface, the boys realized that it was a shark and scrambled for the bank. The shark grabbed Lester and pulled him underwater.

The boys ran for help and several people arrived including a local businessman, twenty-four-year-old Stanley Fisher. Fisher and several others waded into the creek to try and find Lester. They thought he had suffered a seizure. Fisher soon located the boy's body.

But then they all witnessed the shark savaging Fisher and biting into his right thigh. He bled to death in Monmouth Memorial Hospital, Long Branch, a couple of hours later.

Later that day, Joseph Dunn, fourteen, of New York City was attacked a half-mile from the Wyckoff Dock, nearly thirty minutes after the fatal attacks on Stillwell and Fisher. The shark bit his left leg, but Dunn was rescued by his brother and friend after a vicious tug-of-war battle with the shark. He made a full recovery.

Lester's body was found 150 feet upstream two days later.

In an idiotic statement, James M. Meehan, former director of the Philadelphia Aquarium, told the *Philadelphia Public Ledger*:

Despite the death of Charles Vansant and the report of two sharks having been caught in that vicinity recently, I do not believe there is any reason why people should hesitate to go in swimming at the beaches for fear of man-eaters. The information in regard to the sharks is indefinite and I hardly believe that Vansant was bitten by a man-eater. Vansant was in the surf playing with a dog and it may be that a small shark had drifted in at high water, and was marooned by the tide. Being unable to move quickly and without food, he had come in to bite the dog and snapped at the man in passing.

On July 14, Michael Schleisser, a taxidermist and lion tamer for Barnum and Bailey's Circus caught a seven-foot, six-inch white shark in Raritan Bay, a few miles from the mouth of Matawan Creek. The shark nearly sank the boat before Schleisser killed it with a broken oar. Scientists later identified its stomach contents as human flesh. The carcass was displayed in the window of a New York shop, but what happened to it later was unknown.

Was this fish the Matawan killer? Great whites cannot survive in fresh water for any length of time. This shark may be the one that killed Vansant and Bruder, but the one that killed Stillwell and Fisher would have been a bull shark.

Marcia Hathaway was a thirty-two-year-old actress and had played in music halls with touring troupes as well as radio serials. She landed a leading role in a movie, *The Shadow of the Boomerang*, produced by World Wide Pictures.

On January 28, 1963, she had her career cut short whilst wading Sugarloaf Bay in Middle Harbour, Sydney, New South Wales, Australia. It was only two days after Australia Day, and the area was still full of holiday makers. Marcia, her fiancé Frederick Knight, and four friends arrived at Sugarloaf Bay aboard the cabin cruiser *Valeeta* for a picnic. Two of the group were on the rocks, gathering oysters, and others were on the beach when Hathaway and Knight entered the water. Hathaway was bitten, at first she thought by an octopus, but then the assailant bit again, harder and deeper. The actress had been bitten on the buttocks, thigh, and calf.

The others rushed to help, pulling her away from the shark and getting her onto the boat. The cruiser went to the nearest boat shed. Knight dived into the water and swam to the closest house, where he telephoned for help, and an ambulance was waiting at the jetty when the boat arrived. The ambulance burned out its clutch climbing the hill, despite the efforts of volunteers to push it over the rise. A second ambulance was called, but arrived too late. She died en route to Mater Misericordiae Hospital in North Sydney from blood loss and shock.

Most researchers think a bull shark was involved.

Beau Martin was only twenty-three years old, but he was already a doctor of philosophy, youth ambassador, and humanitarian. On December 16, 2002, he went for a swim at Lake Heron, Queensland, Australia, with his friend David Dadd. He later said, "We swam to the other side of the lake and were on our way back, talking and just doing breaststroke. I heard Beau's cries for help and I swam back out but I couldn't find him. He had vanished."

An extensive search failed to find Beau. But two days later, his father found his body. He was missing his left leg. A postmortem revealed he had been bitten three times by a bull shark.

Friedrich Burgstaller, a retired Austrian politician, was on holiday in South Africa in 2014. He was with a fifteen-strong tour group at Second Beach, Port St. Johns, Eastern Cape Province.

The local council was no longer paying for lifeguards, and beach signs had been erected warning people against going into the water. A local woman, Catherine Yazbek, has a guest house by the beach and said many people, including the Austrian man, had ignored the warning signs.

Burgstaller and his wife went swimming, and off-duty lifeguards Sizwe Dusubana, Siphosoxolo Njila, and William James saw the attack. They saw the man being pushed into the air as the shark bit into the lower part of his body. Burgstaller then tried to run, but fell.

He then turned around to face the shark and tried to hit it on the head, but its jaws were open, and his arm fell right between its teeth and the arm was gone, Said Dusubana.

He collapsed and the current started dragging him out to sea. "His head was bobbing in and out of the water, and you could see that he was still alive and fighting," said Njila.

They estimated that the man was alive for about twenty minutes after the attack, trying to swim to shore despite one arm being bitten off.

They tried to phone for a boat, to no avail, then tried stopping passing cars to see if anybody had jet skis or boats, but nobody did. About an hour later, they managed to recover the body after it had floated near a group of rocks that stretched out into the sea. His arm and a large part of his lower body had been eaten.

Geremy Cliff of the KwaZulu-Natal Sharks Board examined the body in the Lusikisiki Mortuary and said that he suspected that a single bull shark around six foot six inches long had been involved.

Stella Berry, sixteen, was a schoolgirl from Cottesloe, Perth, Australia. She was relaxing by a rope swing with her friends in the Swan River in North Fremantle, at about 3:30 on Saturday, when she saw a pod of dolphins and jumped into the water to swim with them. Moments later, a large bull shark latched into her leg.

A man dived into the river to help her and managed to drag the girl to shore, but paramedics were unable to save her.

Ironically, Stella had been featured in a local paper in 2017 for her design of a shark-free ocean pool for Perth's Cottesloe Beach, as part of a school engineering project.

Stella, who was in Year 5 at St Hilda's Anglican School for Girls at the time, worked with one of her classmates to come up with a design for the ocean pool. They were praised for their ingenuity by then-Cottesloe mayor Jo Dawkins. Ms. Dawkins told *Perth Now* she was impressed with the design because it allowed smaller fish to enter the pool, while keeping predatory creatures out.

THE OCEANIC WHITETIP

A pelagic (open ocean) shark, the oceanic whitetip (*Carcharhinus longimanus*) is found in tropical and semi-tropical seas worldwide. It averages 9.8 feet long, but can run up to 13 feet. Its notable characteristics are its long pectoral fins, which resemble the wings of a plane.

Due to the dearth of prey in the open oceans, the oceanic whitetip is known to be an aggressive opportunist and has little fear of man. It feeds on sea birds, fish, marine mammals, and carcasses. In its scant environment, it must take every opportunity it can to feed. It has been known to attack, kill, and eat humans.

The most famous attacks on people were mass attacks on the crews of sunken ships in World War II. The *USS Indianapolis* was a *Portland*-class heavy cruiser. In July of 1945, she was sent on a top-secret mission to deliver uranium and other components for "Little Boy," the first nuclear weapon used in combat. She was torpedoed by the Japanese sub 1-58 on July 30 in the Philippines Sea, causing the greatest loss of life from a single ship in the US Navy. Of 1,195 crewmen, only 316 survived the ordeal. The ship took only twelve minutes to sink, and many men in the water had no life rafts or life jackets.

The survivors were afflicted with dehydration, hypothermia, skin damage due to exposure to sea water, jellyfish stings, and shark attacks.

It is thought that the main culprits were oceanic white tips. It is thought that 150 men were taken by sharks, and doubtless the sharks also fed on those already dead.

Also during WWII, the Royal Mail Ship *Nova Scotia* was sunk in 1942 by a German sub in the Indian Ocean. The RMS *Nova Scotia* lost 858 of the 1,052 people aboard, including 650 Italian internees, 96 crew members, 88 South African guards, 10 DEMS gunners, eight military and naval personnel, five passengers, and the *Nova Scotia*'s master. The U-177 sank her with three torpedoes. Many of those who ended up in the sea were taken by sharks, most likely oceanic whitetips.

On October 14, 1972, Rodney Temple, together with Bret Gilliam and Robbie McIlvaine, were diving off Cane Bay, Saint Croix, US Virgin Islands. At a depth of 210 feet, they saw two oceanic whitetips circling them. The men began to ascend, stopping to decompress on a reef ledge at 175 feet. Gilliam and McIlvaine were above Temple and noticed bubbles rising up from below. Diving down, they found one of the sharks biting into Temple's thigh. Soon after, the second shark savaged his leg, severing the lower portion.

Temple's friends attempted to pull him up, but the sharks began savaging his torso and dragged him down to 400 feet. Gilliam and McIlvaine surfaced and survived, but Temple's body was never recovered. Gilliam had to be evacuated to Puerto Rico to be treated for decompression sickness. He was cited for heroism by the Virgin Islands governor. Gilliam later became a pioneering technical diver and coauthored over seventy-five books.

On June 30, 1974, Edward Horne, a Houston real estate executive, and his family sailed from Panama City, Florida, aboard the forty-three-foot motor yacht *Princess Dianne* bound for the Bahamas. On July 1, the Coast Guard got a mayday call saying that she was taking on water, but they were unable to find the yacht in the darkness. Shortly before noon on Tuesday, a plane spotted the cabin top and wreckage from the Princess Dianne.

The pilot saw the seven family members hanging onto a ring-shaped life preserver whilst sharks appeared to be feeding on one of them. The older family members were trying to kick the sharks away. The pilot discovered a small pleasure craft, the *Mabel II*, nearby, signalled "follow me," and directed them to the shipwrecked family. The family was taken aboard the pleasure craft until the Coast Guard cutter *Point Lobos* arrived. When rescued, the family had been in the water for twelve to thirteen hours.

Billy Horne, age ten, had been bitten on the arm and shoulder; his three-year-old brother Edward Horne Jr. had died of exposure. Billy was picked up by a helicopter diverted from another mission. The helicopter was too low on fuel to bring him to Tyndall Air Force Base in Florida. One of the crewmen gave him mouth-to-mouth resuscitation, but the helicopter crew believe he expired about halfway between the boat and land.

The attacking sharks were thought to be oceanic whitetips some twelve feet long.

In 2010, at the Egyptian Red Sea resort of Sharm el-Sheikh, an oceanic whitetip shark savaged four bathers, killing one of them. The German tourist lost an arm in the attack. The authorities closed the beaches until the shark thought to be involved was caught. The open-water fish had been far closer to shore than usual. Environmentalists theorized that overfishing and a declining ecosystem drove sharks closer to shore in search of food. There are also accusations that tourist boats were illegally dumping meat into the water to attract sharks for passengers wanting to photograph them. It was also said that the crew of a ship transporting livestock had dumped dead animals overboard.

A handful of other species, such as the short-fin mako (*Isurus oxyrinchus*), the blue shark (Prionace glauca), and the bronze whaler (*Carcharhinus brachyurus*), have killed a few people, but these attacks were usually provoked and not predatory attacks.

THE CAUSES OF SHARK ATTACKS

Humans are not the natural prey of sharks. Unlike other beasts in this book, our worlds do not usually intersect. No shark comes out onto land to hunt people. In the case of the great white shark, there is a theory that some attacks are cases of mistaken identity. Great whites usually feed on marine mammals and attack them from below. When a person is on a surfboard, the outline they present to the shark is very like that of a seal or sea lion. The shark bites the surfer, mistaking it for a fat-rich marine mammal, only to get a mouthful of surfboard and bony, bad-tasting human flesh. Some shark attacks, like the one on Louis Boren, seem to be of this nature—a single fatal bite.

It is also thought that great whites explore things with their mouths. Lacking hands, sharks mouth strange objects. An exploratory bite from a big shark can prove fatal to a human.

Some great white attacks end up with the victim getting consumed. It may all be down to how hungry the individual shark is. In the Mediterranean, great whites almost always eat their human victims. This may be due to a lack of large seal populations in the area. Great whites feed mainly on dolphin and tuna there, so maybe they are a bit less picky. What is without doubt is that great whites prefer calorific, fat-rich prey over bony human titbits.

Bull sharks often hunt in murky water and rely on smell and electroreception more than sight. If a person is swimming in murky water, especially if there are bait fish about, the shark may mistake them for its more normal prey.

The tiger and the oceanic white tips tend to be generalists, the former often called the dustbins of the sea as they consume almost anything they can.

When animal carcasses or fish guts are tossed into the water, they may attract sharks that will then associate humans with food. Likewise, spearfishing can release blood into the water that can attract sharks.

Another factor is overfishing by man. Humans can deplete areas of fish, leaving predators starving. This is what many experts think caused the attacks in the Red Sea in 2001, 2022, and 2023.

HOW TO AVOID SHARK ATTACKS

The best way to avoid shark attacks is simply not to swim in waters that they frequent. If you must swim, then follow certain rules. Do not swim at dawn or dusk, when sharks are more active. Do not swim in murky water or where there are shoals of fish. Do not swim near areas where people are gutting fish or where animal remains are in the water. Do not swim near sandbanks or drop-offs, as sharks lurk near these. Do not wear jewellery, as it can catch the light and look like a shiny fish. Humans are clumsy in the water, and there is some evidence that our thrashing around in the sea can sound like an injured or distressed fish.

If you are attacked, then fight back. Sharks usually bite their prey and then retreat as it bleeds out. They try to avoid fights with marine mammals that may scratch and bite, hence the attack-and-retreat strategy. Punch, scratch, and gouge at the shark's sensitive areas, such as the eyes, gills, and snout. This worked for thirteen-year-old Ella Reed, who was attacked by a bull shark off Fort Pierce, Florida in 2003. When attacked by the shark, she bravely punched its face and nose, driving it off twice and allowing other people a chance to rescue her.

People would do well to remember that only a tiny number of humans are killed by sharks each year, an average of five. In contrast, humans kill a staggering one hundred million sharks each year. Killed for their fins, meat, and oil, the market is worth a billion dollars per year. The global population of sharks has dropped by 70 percent. Add to this deaths from pollution and entanglement in nets, and it becomes clear that humans pose a much greater threat to sharks than vice versa.

CHAPTER FOUR

BEARS

"If it's black fight back,
if it's brown lie down,
if it's white goodnight."

—SAYING ON HOW TO REACT TO BEAR ATTACKS

In August 2015, Lance Crosby, sixty-three, of Billings, Montana, was hiking alone in Yellowstone Park. He was off-trail and not carrying bear spray. He was attacked and killed by a female grizzly bear who, together with her cubs, fed on his body before caching it. When Crosby's body was found, he had claw and bite marks on his arms, suggesting he was trying to defend himself from the 259-lb bear.

After his remains were discovered, a female grizzly was caught close by in a bear trap. DNA tests showed that she was the animal involved in the attack, and she was subsequently shot. Her two cubs were sent to a zoo.

There are eight living species of bear. They are large omnivores of the family Ursidae. Some, like the giant panda (*Ailuropoda melanoleuca*), feed mostly on plant matter, mainly bamboo. At the other end of the dietary scale is the polar bear (*Ursus maritimus*), which, due to its habitat, feeds exclusively on flesh. Bears are found in Asia, Europe, North America, and South America. They are absent from Australia, Antarctica, and Africa, though one species, the Atlas Mountains bear (*Ursus arctos crowtheri*), lived in North Africa until historic times. The last one was killed in 1870, and its extinction was due to human hunting.

Bears usually avoid humans, and in many areas have lived alongside them with no attacks. But some bears do kill people. There are two types of bear attack: the defensive attack, where the animal has perceived the human as a threat or rival, and the rarer predatory attack, where the bear kills the human to feed on the carcass.

THE POLAR BEAR

Ursus maritimus vies with the Kodiak bear (*Ursus arctos middendorffi*) for the title of the world's largest bear. It is found throughout the Arctic region and is the supreme land predator of the North Pole.

As with all large predators, the polar bear has left its mark on culture in legend and folklore.

In Eskimo legend, Torngarsuk is the god of the sea, death, and the underworld. He takes the form of a polar bear and is master of whales and seals.

Only the *angakkuit*, the Eskimo shamen, can see Torngarsuk. The shamen worship him with song and dance. They keep familiar spirits in leather bottles and send them to the bear god's cave to ask for healing and good fortune. Polar bears are also linked with the creator goddess, Nuliajuk.

In some stories, people transform themselves into polar bears by donning a polar bear skin, much like the werewolves of Europe, which we will look at in a later chapter.

The polar bear also features in Norse folklore. Bjarndyrakongur was the polar bear king, a bear with red cheeks and a unicorn-like horn, who could understand human language. His horn glowed in the dark!

Polar bears, due to their Arctic home, are almost exclusively carnivorous. They feed on seals, walruses, narwhals, belugas, musk ox, reindeer, fish and their roe, sea birds, and eggs. They will also scavenge whale carcasses. Some plant material, such as seaweed and berries, may be eaten in small amounts.

They are endowed with powerful jaws, sharp teeth, and hooked claws. Their broad paws work like snowshoes. Polar bears can run at 25 miles per hour and swim at 3.7 miles per hour.

The white-looking fur is actually transparent, scattering light between the underfur and the skin for absorption and re-emission.

Males average 8.2 feet long and 1,760 lbs. The largest confirmed specimen was shot at Kotzebue Sound, northwest Alaska, in 1960. It tipped the scales at 2,209 lbs and, when standing up on its hind legs, was eleven feet one inch tall.

Polar bears can and do hunt, kill, and eat humans.

Willem Barentsz (1550–1597) was a Dutch cartographer and Arctic explorer. During his second trip to the Arctic in 1595, he recorded that the party was attacked by a polar bear on Cherie Island.

A party landed upon the island and procured a quantity of eggs; and as they were returning they encountered a large white bear which fought with them a while. Four glasses [hours] ranne out and she swam away with a hatchet that had been struck into her back; the ferocious animal was, however, killed at last and found to be thirteen feet in length.

The attack had left two men dead. The size suggests that the bear was male despite being referred to as "her."

On January 5, 1975, Richard Michael Pernitzky, eighteen, was killed by a polar bear at an oil exploration site in the Mackenzie River delta, Canada. When Pernitzky failed to return to his living quarters, a search found a polar bear eating his remains only nine hundred feet the from camp. The bear was later hunted down and killed and had three metal tags in its ears, indicating it had been removed from civilization three times previously.

Forty-six-year-old Thomas Mutanen of Churchill, Manitoba, was attacked and killed on a city street by a young polar bear in November of 1983. It was one of the bears migrating through the area. A lack of sea ice compelled them to walk along the coast and through the city. Passersby distracted the bear, and it was shot, but Mutanen later died.

On December 8, 1990, Carl Stalker, twenty-eight, of Point Lay, Alaska, was walking with his girlfriend when a polar bear chased him down into the middle of the village and killed him. The bear was three-quarters of the way through eating him when it was shot by Scott Haugen.

At five in the morning on Saturday, a call went out over the radio that the only police officer in town needed help to find a bear that had dragged away a villager.

All that remained in the road where the attack took place were blood and bits of human hair, Haugen said. While villagers on snowmobiles began searching a wide area, Haugen was told by the officer to take his rifle and follow the blood trail.

He tracked the bear's progress about a hundred yards down an embankment towards the lagoon.

I shined a light down there and I could see the snow was just saturated with blood.

A snowmobiler drove up, and in the headlights, Haugen discovered what was left of Stalker. Then, as the lights of another snowmobile reflected off the lake, Haugen saw the hunkered form of the polar bear.

When they hunt, they hunch over and slide along the ice to hide the black area of their eyes and snout. It wasn't being aggressive toward us, but I wasn't going to wait.

Later, wildlife biologists told him that this was the first recorded US case of a polar bear killing a human being.

On the foggy night of August 5, 2011, a group of students from the British Exploring Society were camped near the Von Post glacier, on the island of Svalbard in the Arctic circle. During the night, the group was attacked by a starving polar bear. A warning tripwire system, used to set off flares that might have scared the animal away, failed to go off and it attacked, killed, and partially ate seventeen-year-old Horatio Chapple and badly mauled several others, including Michael Reid, twenty-nine, the expedition leader, and fellow leader Andy Ruck, twenty-seven. Reid was badly bitten on the head before managing to shoot the bear dead with a rifle that misfired four times.

Patrick Flinders, sixteen, and Scott Bennell-Smith, seventeen, who were sharing a tent with Horatio Chapple, were also savaged. Flinders underwent an operation in Norway to remove parts of the bear's teeth from his scalp, while Bennell-Smith suffered a broken jaw and smashed teeth.

The remote community of Wales sits on the coast of Alaska, just fifty miles from Russia across the Bering Strait. It is accessible only by boat or plane that are used to ship in goods. Only 150 people live there and there are no police.

Early in 2023, Summer Myomick, twenty-four, and her one-year-old son Clyde Ongtowasruk were walking the three hundred feet from the village school to the health clinic. A blizzard was blowing, so they never saw the polar bear that stalked them. The beast attacked them just after they left the school and onlookers saw them both get mauled by the massive beast.

School officials rushed people into the building for safety. The headteacher then ordered a lockdown and closed the blinds so the children couldn't see the attack unfold.

Several community members attempted to save the mother and son by attacking the bear with shovels, causing it to stop briefly and turn its attention to them. The bear chased them back into the school, before Principal Dawn Hendrickson slammed the door in the face of the charging bear.

Susan Nedza, chief administrator of the Bering Strait School District, said, "The polar bear was chasing them and tried to get in as well. Just horrific. Something you never think you would ever experience."

A called went out to community members for help, and an unidentified person turned up and managed to shoot the bear dead. But both mother and son were already dead.

It is probably due to the lack of large human populations in its range that this macropredator has not killed more people.

THE BROWN BEAR

The brown bear (*Ursus arctos*) is the most variable of the bear species, with as many as twenty-one subspecies scattered around the world, though five of these are thought now to be extinct. It is also the widest ranging bear, found in parts of North America and once ranging south as far as Mexico, across much of Europe and North Asia as far as Japan and south into Central Asia, parts of the Himalayas, parts of China, and parts of Korea. One subspecies once lived in North Africa.

Throughout its range, it has made a mark on folklore and legend.

Viking berserkers were warriors who wore bearskins. They were thought to gain the strength and savagery of a bear by wearing its skin. They fought in a trance with superhuman fury, foaming at the mouth like wild beasts. The Icelandic historian and poet Snorri Sturluson describes them thus:

> ***His (Odin's) men rushed forwards without armour, were as mad as dogs or wolves, bit their shields, and were strong as bears or wild oxen, and killed people at a blow, but neither fire nor iron told upon them. This was called Berserkergang.***

The seeds of the psychoactive plant henbane (*Hyoscyamus niger*) have been found in a Viking grave that was unearthed near Fyrkat, Denmark in 1977. It has been suggested that the effects of henbane could have caused the berserker rages.

The father of the gods, Odin himself, would sometimes take on the form of a bear when he visited the mortal realm.

Veles, the Slavic god of the underworld, water, and animals took on the form or a dragon or a bear.

Farther east, in Japan, bears have had an influence. Inoshishi is a magical female bear who has the power to drive away snakes. Even reciting her name will banish serpents.

Onikuma was a mountain-dwelling demon bear. It was said to be of great age, size, and strength. This was the giant bear that terrorized Nagano prefecture. It devoured many horses and cattle. The hunter who finally slew it claimed that its hide could cover six tatami mats.

The Ainu, the aboriginal people of Japan, have legends of the *payep kamui*, magical bears that grow from the roots or branches of trees. No mortal hand can kill them.

Native Americans have many stories about the brown bear.

The Nez Perce people of the northwest tell a story of a man named Five Times Surrounded in War who went fishing and hunting up the Grande Ronde River. A female grizzly bear takes on the form of a beautiful girl and asks him to marry her. He falls in love and says yes. She takes him back to her camp, and it is filled with dried meat, fish, and fruit. She lives in a hole in the ground, and he realizes that she is a bear in disguise. She later kills one of a group of hunters, but the others lure her out with a decoy and kill her.

The brown bear vies with the polar bear for title of the largest bear. The Kodiak subspecies (*Ursus arctos middendorffi*) inhabits the Kodiak archipelago in Alaska. The largest recorded specimen was a male called Clyde, who lived at the Dakota Zoo in Bismarck, North Dakota. He weighed a whopping 2,400 lbs.

Being omnivores, brown bears eat a vast array of food that can vary with habitat and time of year. Their diet includes edible fungi, acorns, grasses, berries and other fruits, roots, flowers, and nuts. They also feed on insects, honey, fish (salmon being a particular favourite), and bird's eggs. Bears hunt and kill a wide variety of mammals, from the small, such as chipmunks and marmots, to deer, sheep, goats, moose, and even musk ox and bison. In Siberia, the Kamchatka brown bear (*Ursus arctos beringianus*) has been known to kill and eat full-grown Siberian tigers. Throughout their range, brown bears have, on occasion, killed and eaten humans.

Perhaps the most determined and savage brown bear attacks took place over six days in Sankebetsu Rokusen-sawa, a remote village in Hokkaido, Japan, in November of 1915.

It began one morning in mid-November, when an Ussuri brown bear (*Ursus arctos lasiotus*) turned up at the Ikeda family's house, where it scared their horse and made off with a sack of corn. The bear returned on November 20 and again on November 30, when the head of the family's son Kametaro, and two traditional winter hunters or *matagi*, confronted it and shot it. The bear was only wounded, and the men tracked the blood trail towards Mount Onishika, but a snowstorm forced them back. The men felt the bear would not return to the area. They were very, very wrong.

On December 9, 1915, at 10:30 a.m. the huge bear returned. This time it turned up at the farm of the Ota family. The farmer's wife, Abe Mayu, was at home with a baby that the family was looking after, Hasumi Mikio. The bear killed the baby by biting her head. Mayu tried to drive it off, but it chased her out of the house, caught and killed her, then dragged her into the forest to eat. The farm looked like a slaughterhouse with pools of blood on the floor.

Next day, a thirty-man group set out to kill the bear and recover Mayu's body. They met the bear in the forest and five of them shot at it. Only one shot hit, but it drove the bear away. The men recovered the woman's body. It had been mostly devoured by the bear.

The bear returned to the Ota farm at eight o'clock in the morning. This time the villagers were waiting for it with guns. When it appeared, the villagers panicked and only one actually shot at it.

Fifty guardsmen pursued the bear, thinking it had gone downriver. The beast, however, visited the farm of Miyoke Yasutaro, who was away from home with his friend and neighbour Saito Ishigoro, leaving only the women and children from both families on the farm. Yayo, the farmer's wife, was cooking dinner with her fourth son, Umekichi, on her back. She heard an odd rumbling noise and then the bear broke through the window. It overturned the cooking pot, putting out the fire, and as the family panicked, the oil lamp was doused, plunging the house into darkness.

Yayo tried to run from the house, but her second son, Yujiro, clung to her legs in fear, tipping her over. The bear bit into Umekichi. The one guard at the house, a man named Odo, ran to investigate. The bear released the mother and child, who ran away, and attacked him. Odo tried to hide behind furniture, but the bear clawed his back.

The bear then attacked Kinzo, the third son of the Miyouke family, and Haruyoshi, the fourth son of the Saito family, killing them, and bit Iwao, the third son of the Saito family. Next to be attacked was Take, Saito Ishigoro's pregnant wife. She too was attacked, killed, and eaten. From later testimony, villagers heard Take begging the bear not to touch her belly but instead to eat her head.

The guards, realizing they were not on the killer bear's trail, returned to the village. A badly injured Yayo told them that the bear was at the Miyouke family's house. When they arrived, the house was in darkness and the guards thought the bear had killed all inside. They wanted to set fire to the farm, but Yayo thought there might be survivors inside.

The guards broke into two groups. One stood at the front door, and the second group went to the rear to set up a racket in the hope to drive the killer out of the front of the house into a hail of fire. But the plan failed, as the bear emerged from the front door. The guards, however, had bunched up in such a way that the men were in each other's line of fire. Only the guard at the front had a clear shot, and his gun misfired. The bear escaped into the night.

Only Rikizo and Hisano, first son and daughter of the Miyouke family, survived the attack. They were moved to the Tsuji family house near the river. The events spooked the guards so much that most left, leaving only the brave veterans of the Russian-Japanese war to stay on protect the villagers.

In the meantime, Saito Ishigoro, unaware of the horror that had befallen his family, filed a report with authorities and the district police before returning to Tomakomai and lodging at a local hotel.

Miyouke Yasutaro had heard that a man named Yamamoto Heikichi was a veteran bear hunter. He visited Heikichi's house, and the old man told him that he thought that the bear was one he had named "Kesagake" or sash, due to the sash-like markings of the fur around his shoulder. Kesagake had previously killed and eaten three women, according to the bear hunter. However, the old man had retired and had pawned his gun for money to buy alcohol, and he refused Miyoke's request to hunt the beast.

On December 11, Miyouke Yasutaro and Saito Ishigoro returned to Sankebetsu and found out about the horrific events. They formed a group of men to hunt and kill Kesagake. Thinking that he would return to Miyouke's house, they waited there that night, but the bear did not return.

The events reached the ears of the Hokkaido Government Office, and a sniper team was organized under the Hoboro branch police station and led by Chief Inspector Suga. The team went to Sankebetsu on December 12. Once again Kesagake did not appear.

At dawn the following day, the bear ransacked the Ota family house and ate their stock of winter food. Suga had his men search the surrounding hills to no avail. He created an ice bridge across the river and set snipers on it. At night one man saw a shadow near some tree stumps and the snipers opened fire, but the shadow vanished.

However, the next morning they found bear tracks and blood, indicating that Kesagake had been wounded. Bear hunter Yamamoto Heikichi had a change of heart and travelled up to Sankebetsu to hunt his old foe. He set out with local guide Ikeda Kamejiro and tracked Kesagake. The pair found him asleep under a Japanese oak. Heikichi shot him through the heart, killing him instantly.

Heikichi was measured at eight feet nine inches tall. He weighed 750 lbs, very light for a bear of that size. In an autopsy, human remains were found in his stomach. His skull and hide were kept but have since been lost.

Yayo made a full recovery, but Miyouke Umekichi, who was bitten by the bear while being carried on his mother's back, died less than three years later from the wounds.

Odo recovered from injury and returned to work, but the next spring, he fell into a river and died.

The village of Sankebetsu was abandoned and became a ghost town. A shrine and a statue of Kesagake stand there today.

Kesagake himself was a man-made monster. Deforestation and hunting of prey species made hibernation impossible for him due to lack of food. He approached human habitation in search of food because humans had taken natural food from him. He became a man-eater of necessity.

On June 12, 1922, Joseph "Frenchy" B. Duret, a pioneer, hunter, trapper, and guide was in the Absaroka Beartooth Wilderness, Montana. He had trapped a huge grizzly bear, but the animal broke out of the trap.

Duret shot the beast, but the bullet did not hit a vital spot and the bear mauled, killed, and ate him.

John MacDonald, a seventy-year-old woodcutter who lived alone in a cabin on the Yukon River, was attacked, killed, and partially eaten by a grizzly in October 1932. His remains were found in the bush at a place called Campbell Slough. They were taken to a cabin and the door securely barred. But the bear returned to finish its meal. Breaking down the door, the bear devoured what was left of the man and scattered his bones in the bushes outside. The bear was never caught, and it was thought that it had turned man-eater due to the failure of berries that year.

In 1959, laboratory studies on the remains of Sam Adams concluded that he had been killed and eaten by a grizzly bear the previous year. Sam, forty-eight, had been hunting in Montana, seemingly the capital of grizzly attacks, with two friends, Ed Hoskings and Calvin Trusky, when they split up. Sam never returned. The two friends led a search party that found his remains on a mountain meadow as the snows thawed. Heavy snow had prevented an earlier search.

Sam's bones were scattered over an area of fifty feet and his boots had been gnawed. His rifle was smashed. Special Deputy Charles Schimiedeke said:

> ***It looks like Adams wounded the bear. He used all but two cartridges in the battle, but was hurt. He apparently crawled alongside a log and tried to light a fire to get warm. He removed shoelaces probably to use as tourniquets. Then it looks like he fired his last two bullets to summon aid.***
>
> ***Then the bear reappeared and Adams used his gun as a club. The evidence is that Adams was killed and eaten in a wild battle with a bear.***

He went on to say that the bear had devoured Adams on the spot.

August 13, 1967, was a date that would live in infamy and become known as "The Night of the Grizzlies." Two campers were killed on the same night in the same park, Glacier National Park, Montana.

In the wee small hours, a grizzly bear dragged nineteen-year-old Julie Helgeson from her tent at Granite Park Chalet and began to maul her. In the fifty-seven years of the park's history, there had never been a bear attack. The bear also bit Helgeson's boyfriend, Roy Ducat, on the arms and legs before returning to the girl. Ducat ran for help and came back with a group of campers. The bear had dragged its victim four hundred feet into the woods. She was rescued alive, but badly mauled. She died of blood loss on a makeshift operating table at the chalet at 4:12 a.m.

At the same time, just nine miles away, another grizzly bear attacked a group of five campers at Trout Lake Chalet. Most of them managed to flee and climb trees, but nineteen-year-old Michele Koons could not unzip her sleeping bag. The bear grabbed her, sleeping bag and all, and carried her off three hundred feet into the forest.

Koons's four friends remained in the trees until dawn, then ran for the nearest ranger station.

Subsequently, both bears were shot after being baited out. Both were females and were underweight. It was found that in other areas, the chalets had been feeding food waste to the bears, which then lost their fear of people and began to associate humans with food. Then it was only a short jump to looking at humans as food.

The body of forty-eight-year-old Henry Cardinal was found in dense bush near Fort St John, British Columbia, on January 14, 1970. The corpse had been partially eaten. Tracks in the snow indicated that Cardinal had been stalked, killed, and eaten by a grizzly. Rangers tracked the beast by helicopter and shot it. It was found to have a gum infection that may have prevented it from hunting its normal prey.

In the summer of 1970, five members of the Fukuoka University Hiking Club were spending eleven days hiking across the Hidaka mountain range in Hokkaido, Japan. On the evening of July 25, they were getting ready to ascend a mountain from which they hoped to see the sunrise the next day. As they were finishing dinner, a small Ussuri brown bear entered the camp. The young bear tried to make off with the packs containing the hikers' food, so they attempted to scare it away by banging pots and pans. The bear ran off, and later the campers entered their tents to sleep.

At nine o'clock that night the bear returned and ripped into one of the tents. Again, they scared it off by banging utensils, and then lit a fire. They decided to take turns keeping watch. The bear did not return and finally morning came. Instead of returning back down the mountain, they thought that they would press on to the summit. The group had come all the way from southern Japan, and they would be damned if they were going to let a small bear stop them from their goal.

But as they were breaking camp, the bear returned and was acting in an aggressive manner. They all hid in one of the tents, but the bear began to tear it down. The group escaped from the rear of the tent. The animal rummaged through their packs.

The hikers decided to split up. Three of them would try to salvage what was left in their camp once the bear had left. Two others would head down the mountain to get help.

The two running downwards met with another group of hikers from a local university. These students had also been attacked by the same bear and advised the pair to head down the mountain with them to safety. However, they declined, not wanting to leave the others behind. The students from the local university promised to contact the authorities once they were off the mountain.

Climbing back up the mountain they met with their three friends, who had managed to rescue their tent and some of their food supplies. As the light was fading, they decided to make camp in a new area, hoping that the bear would not come back. In the morning, they planned to descend.

As they made their new camp, the bear appeared yet again, having tracked them. In a panic, they ran down the mountainside even though the sun was setting. The bear pursued and quickly caught one of the youths, dragging him off and killing him. The others heard his screams.

In the blind panic, another of the group had got separated from his friends. He stumbled on a campsite left by the other hikers. This boy wrote a memo of what had happened that night. The journal was later found by a search party. It read...

I [threw a] rock at the bear and hit its snout. I think it stopped following me after that. Found an abandoned camp, I'll spend the night here. I wonder what happened to the others.

He continued...

I can't sleep. The slightest sound of the wind rustling the grass scares me. I hope rescue is coming soon. The sun is coming up but I think I'll stay inside the tent until it get[s] a bit brighter. I want to go home.

Then a little later...

Oh my god the bear is outside. Have the other members made it off the mountain? When is rescue coming? I am afraid. The fog is getting thicker.

The search party that found his journal also found his body, torn and devoured.

The remaining three were walking down the mountain through the thickening fog. Suddenly the bear exploded from the fog and attacked. In a selfless act of bravery, the oldest youth fought the bear, allowing the two others to escape.

Around noon, the two survivors stumbled onto a group of construction workers and told them what had happened. They contacted the authorities, who had already been alerted by the local hikers.

The rescue party found the other three members dead and partially eaten. They also found and shot the bear that had killed the boys. Its stuffed body is still on display at the Hidaka Hiking Centre. The bear is a young one, no larger than a big dog. This case goes to show that even small bears can become killers.

In Izembek National Wildlife Refuge, Alaska, a nature photographer, Jay B. L. Reeves, was attacked, killed, and eaten by a grizzly bear on August 3, 1974. A fisherman stumbled upon his camp and found it had been ravaged by a bear. He alerted rangers and a helicopter was dispatched. The crew saw a grizzly emerge from a patch of alder trees four hundred feet from Reeves's camp.

The bear was shot and found to have human remains in its stomach. Reeve's bones and skull were found close by. The bear was perfectly healthy and about five years old.

On September 10, 1976, twenty-five-year-old Alan Precup arranged a backpacking trip in Glacier National Park, Alaska. He had arranged to meet officials from the visitor's office on September 13, but that date came and went.

On September 16, a group of hikers who were visiting the White Thunder Ridge area of the preserve were asked to keep a lookout for Precup. Their camp was visited by an aggressive grizzly that forced them to retreat into the nearby hills as it trashed their camp. Then, instead of eating the food in the camp, it picked up their scent and followed them. As they moved higher, it stalked them, getting as close as twenty feet away. They drove it away temporarily by throwing rocks at it, but it came after them again. A passing plane distracted it long enough for them to reach the camp of some other hikers and then get to a ranger's boat and raise the alarm.

Park ranger James Luthy investigated with a helicopter crew. They found Precup's destroyed camp. On investigation, he found that everything with human smell on it in the camp had been bitten. The bear appeared but was driven off when the rangers pelted it with rocks. Alan's remains were found close by. The flesh from most of his bones had been eaten except for his right hand that lay under the page of a book and his feet that were still in his socks and boots. His ribs had been scattered across the area.

The bear was never caught, and authorities said that there was an abundance of berries and salmon that year. Like the bear that killed and ate Jay B. L. Reeves, this seems to have been a predatory attack by a healthy, well-fed bear that just looked on the human as food.

Twenty-two-year-old Mary Pat Mahoney was camping at Glacier National Park, Montana, on September 23, 1976. At around seven in the morning, she was dragged from her tent, killed, and partially eaten by a grizzly. An hour and a half later, two young grizzlies were shot. Blood on the claws of one of them was found to be human, and the size and shape of its teeth matched the wounds on the victim's body.

Rangers identified the pair of bears as ones that had aggressively approached campers and fishermen in search of food in past months. In one case, they chased two men into a lake, and in another chased hikers until they dropped their packs, and then fed on the food.

Four years later, in the same park, another grizzly killed and ate two teenagers. Kim Eberly and Jane Ammerman, both nineteen, had been camping illegally in the park. Rangers found their destroyed camp and devoured bodies. Blackfoot tribe members hunted and shot a grizzly whose fur was coated in blood. The bear, according to Clyde Lockwood, the park's head naturalist, seemed to be the same one that had been visiting a dump site at the town of Saint Mary.

The tale is a cautionary one. You should never camp in a national park without telling the authorities first.

Forty-year-old Harley Seivenpiper was hunting alone in Port Alexander, Alaska, on November 4, 1988, when he was killed and dragged a mile to a cache by a grizzly. This shows that it was a predatory attack rather than a defensive one. The bear saw him as food, killed him, then stored him for later consumption. When searchers approached, the bear charged them and was shot.

The ironically named Anton Bear, age six, was walking along a road with his mother and sister near King Cove, Alaska, on July 10, 1992. Suddenly, a grizzly that had been feeding on the local dump approached them. The family made a fatal mistake and ran, triggering the bear's attack instinct. The bear caught Anton and ate him. Two villagers later shot the bear.

British tourist Trevor Percy-Lancaster, forty-six, from Winchester, Hampshire, sacrificed himself to save his wife Cherry when the pair were attacked by a grizzly bear in Jasper National Park, Alberta, Canada, in September of 1992.

When pitching camp in a remote area, they disturbed a 312-lb bear beside a stream. Again, the cardinal rule was broken, and the tourists fled, pursued by the bear. Cherry climbed a tree, but the grizzly bit into her hiking boot and dragged her down. Trevor distracted the bear to save his wife, who then fled. The bear turned on her husband, killing him and eating his leg.

Cherry made a full recovery after a stint in hospital, thanks to her husband's bravery.

A poor berry harvest was blamed for the attack.

May 17, 1998, was the last time Craig Dahl, twenty-six, was seen alive. He had gone hiking alone in the Two Medicine area of Glacier National Park, Montana. His partially eaten body was found three days later and the attack was attributed to a female grizzly and her two cubs. The bears were later shot.

The female was nicknamed "Chocolate Legs" for her distinctive colouring, and had been identified as a problem bear in 1983, captured, and moved to the park's back country.

The next major bear attack in Alaska occurred in 2003, and this one grabbed headlines around the world. Sometimes human stupidity just amazes, and this is one of those times. The story involved a man who thought he could treat grizzly bears like dogs and ended up paying with not only his life, but that of his girlfriend as well.

Timothy Treadwell of Long Island, New York, had exhibited some strange behaviour as a student, claiming to be British or Australian, then changing his second name from Dexter to Treadwell for no apparent reason. After losing the acting role of Woody Boyd in *Cheers* to acclaimed actor Woody Harrelson, Treadwell became an alcoholic. Later, he began travelling to Alaska to study grizzly bears and spent thirteen summers there, until the incident that ended his life.

Treadwell would camp in an area called Hallo Bay, where vast quantities of sedge grass, along with clam-rich tidal flats, attracted large numbers of grizzlies. He would get close to the bears and pet them like big dogs and play with the cubs. It was a disaster waiting to happen.

Tom Smith, a research ecologist with the Alaska Science Center of the US Geological Survey, said that "He was breaking every park rule that there was, in terms of distance to the bears, harassing wildlife, and interfering with natural processes. Right off the bat, his personal mission was at odds with the park service. He had been warned repeatedly."

The National Parks Service kept a file on Treadwell and his multiple violations of park safety rules. These included guiding tourists without a license, camping in the same area longer than the NPS's seven-day limit, improper food storage, wildlife harassment, and conflicts with visitors and their guides. He also refused to carry bear spray or erect an electric fence around his camp.

During the latter weeks of summer, he would move camp to Kaflia Bay and an area of dense bush he called the "Grizzly Maze" due to it being full of bear trails. He seemed to think that by being in the area, he discouraged poachers. He often took his girlfriend, Annie Huguenard, with him.

In October of 2003, Treadwell and Huguenard were camped at Katmai National Park and Preserve on the Alaska Peninsula. Huguenard had been keeping a diary, and later examination reveal that she had been scared of the bears and wanted to leave the park. They were in the park later than they usually stayed. That late in the season, bears are feeding furiously to lay down fat for winter. Apparently, the pair had stayed on an extra week after an argument with an airline over the price of a return ticket. They were meant to leave on September 29.

On October 6, Willy Fulton, an air taxi operator, arrived to pick the pair up from their camp. On the way up the hill to the camp he called their names, but nobody answered. He began to get an eerie feeling.

"About halfway up, I got kind of an odd feeling, and decided to go back to the plane."

That diversion probably saved his life. Looking back over his shoulder, he saw a bear silently stalking him and only twenty feet away. He leapt into the plane and slammed the door.

> ***I've been charged by a few bears, but this was different. He wasn't acting big and bad. He was crouched down, sneaking up on me. That look in his eye was different too. Right then I felt like he was out to kill me and eat me.***

He took off quickly, and flying over the camp, he found it in ruins, with a big bear chewing on what looked like part of a human ribcage.

Fulton contacted the rangers. Four men, including Fulton, flew out to the camp and discovered what little was left of Treadwell and Huguenard. Treadwell's disfigured head, partial spine, and right forearm and hand, with his wristwatch still on, were recovered a short distance from the camp. Huguenard's partial remains were found next to the torn and destroyed tents, partially buried in a mound of twigs and soil. Her head and face were, in a horrid twist, untouched and even looked peaceful, despite most of her body having been eaten.

Suddenly, a large bear appeared and began to defend his prey. Fulton recalled, "He had that same look in his eye, I think he meant to kill all of us."

The rangers all fired at once as the bear began to move towards them, dropping the predator in its tracks.

State troopers Alan Jones and Chris Hill had just landed and heard the commotion. They rushed to the camp area.

As the bodies were being put into bags, a second, smaller bear appeared and was shot by the rangers. A later autopsy found human remains in the stomach of the bigger bear, but not the smaller one.

Trooper Hill had found a digital camcorder whilst looking through the victims' effects. Playing back the contents, Hill came upon a chilling audio sequence of Treadwell and Huguenard's last moments.

Treadwell screams, "Come out here! I'm being killed out here." Huguenard grabs the camera and switches it on whilst screaming to him to play dead. But playing dead didn't work. On the six-minute tape, Treadwell is screaming for three-quarters of the time. The bear finally drags him off into the bush. Huguenard begins to scream and scream. Larry Van Daele, a wildlife biologist for the Alaska Department of Fish and Game, thought that her scream caused the bear to return and kill her too.

The camera's lens cap was left on during the attack, leaving only audio, which is perhaps a mercy.

Bear biologist Matthias Breiter had been camped out some fifteen miles away at the time of the attack. He noted that the berry crop had been very poor that year, and the bears were competing for food. In a stretch of river half a mile long, sixty bears were competing for salmon, when usually there would be only fifteen or so.

The lesson here is never to treat wild animals as pets. It says something for the patience of the bears that they had not already killed him in the previous thirteen years.

On the night of July 28, 2010, a grizzly bear and her three cubs went on a rampage on the Soda Butte Campground at Gallatin National Forest, Montana.

Deb Freele was attacked in her tent. The bear bit into her arm in the middle of the night. She screamed for help, but none came at first. She played dead as the bear mauled her. Then car headlights illuminated the area and the bear retreated. It was a family from a neighbouring camp who said they would drive and get help.

Freele lay alone in the dark for twenty minutes, fearing that the bear would return. The family were too scared to leave the car, so they drove around the campsite's upper loop, honking the horn and trying to rouse somebody. Most people ignored them, suspecting drunks or hooligans. Finally, they roused others to help the stricken woman.

But that had not been the first attack of the night. Shortly before, the same bear had torn into the tent of Ronald Singer, a twenty-one-year-old former high school wrestler, and his girlfriend. Singer drove it off by hitting its snout with a tennis racket but not before it had bitten his leg. His girlfriend's parents drove him to Cooke City, for medical help.

The bear and her cubs moved downstream until she hit the camp of forty-eight-year-old Kevin Kammer. She dragged him out of his tent and, as he was in an isolated area, nobody heard a thing. He was killed and eaten by the bear and her cubs.

The bear was later trapped and killed. She was found to be underweight. The three cubs were sent to Zoo Montana.

In July of 2008 in Siberia's far eastern region of Kamchatka, thirty brown bears trapped a group of geologists at a remote site. The animals surrounded a platinum mining company and killed and ate two guards. Attempts at rescue via helicopter were thwarted by bad weather, and four hundred workers at the plant refused to return.

The poaching of salmon, the bear's main food, in the area, was thought to be the reason for the attacks.

Arrogance can be fatal in the wild, as for instance when somebody wantonly ignores the advice of rangers. Such was the case of fifty-nine-year-old John Wallace, who went hiking in Yellowstone National Park. Rangers had tried to give him advice on bear safety, but he had dismissed them, claiming to be a "grizzly bear expert."

Wallace set off for a hike on August 25, 2011, alone and without bear spray. He was killed and eaten by a mother grizzly bear and her cubs as he sat on a log to eat his lunch. The bear had previously killed another hiker, Brian Matayoshi. That had been a defensive attack, and the hiker's body was not eaten. But it now seemed that the bear was aware of how easy humans were to kill.

Passersby spotted an unattended skiff on Chichagof Island, Alaska in October 2012. On investigation, they found a grizzly bear and her two cubs. A later investigation by state troopers discovered a campsite and evidence of a struggle. They later found the corpse of Tomas Puerta, fifty-four, cached and partially eaten.

The troopers shot the bear, and later found White's camera, showing that he had been photographing the beast from about 150 feet away, much too close.

A bear gained access to the house of Mrs. Claudia Huber near Teslin, Yukon, on October 14, 2014. It chased the forty-two-year-old woman and her husband, Matthias, outside. The grizzly managed to catch and kill Claudia, but her husband retrieved a gun and shot it dead.

An autopsy found that the bear was a twenty-five-year-old male that was healthy but did not have much body fat. The attack was a predatory one. Huber's home did not have any attractants, like an open garage or a barbecue on the deck, which could have lured the bear, so it was targeting human prey.

Lance Crosby, sixty-three, vanished whilst hiking in Yellowstone National Park on August 7, 2015. A worker at a medical clinic, he was reported missing after not turning up to work. A ranger later found his half-eaten body covered in soil and vegetation. The arms bore bite marks from a grizzly, showing that the man had tried to defend himself when the bear attacked.

The bear responsible, a female, was captured. It was found to be the one that had also killed and eaten John Wallace, the self-proclaimed "grizzly bear expert." It had also been involved in the fatal but nonpredatory attack on Brian Matayoshi, also in 2011. The bear was killed, and her cubs sent to a wildlife refuge.

Michael Soltis, forty-four, went missing during a hike along Eagle River, Alaska, in July of 2018. Searchers found his body being guarded by a grizzly bear. The bear attacked and savaged one of the searchers before escaping.

Whilst working at remote Greens Creek Mine on Admiralty Island, Alaska, eighteen-year-old Anthony David Montoya was killed and devoured by a female grizzly and her two cubs. State troopers killed all three animals. Montoya had been operating a drill in an isolated area when he was attacked.

Brown bear attacks in Europe are very rare in comparison to those in North America, but they do occur. In 2019, three men were killed in a little over a month by brown bears in a spate of attacks in Romania. A sixty-one-year-old fisherman was killed in Mures County in October. In late October, a man aged forty-six was fatally attacked in Bacau County. Finally, a sixty-three-year-old shepherd was killed by a bear on the edge of a forest in Mures County in central Transylvania.

Kunshir Island lies in the Kuril Island chain governed by Russia, but just north of the Japanese island of Hokkaido. The evening of October 22, 2019, soldier Alexey Ivanovsky and several of his friends were out digging for crabs on a beach on the island. The friends often collected and ate crabs in their time off from military duties.

The activity caught the attention of a female Ussuri brown bear and her two cubs. The bear attacked, biting into Ivanovsky's leg and dragging him off. She then began to bite at the back of his head, tearing off the skin. She literally skinned her victim alive, ripping the hide not only from his head and neck but from his back and his buttocks. She then began to eat his right leg.

Ivanovsky's friends tried to drive the bear off by hurling rocks, but this didn't work. In the end, they ran back to their vehicle and used the headlights to scare the bear off.

The victim was near dead by the time rescuers arrived in a helicopter, but he was resuscitated and flown 250 miles to regional capital Yuzhno-Sakhalinsk, where he underwent extensive surgery, but died nine days later of a heart attack.

Fifty-four-year-old Toshihiro Nishikawa went on a lone fishing trip at Lake Shumarinai, Hokkaido, Japan in May 2023. A friend who was boating in the area tried calling his mobile phone and got no answer. Later, boat operators saw a large Ussuri brown bear carrying some waders in its mouth. Searchers found a human head in the lake, and it was believed to be that of Nishikawa. A bear was later shot.

As I was writing this book, news broke of a fatal predatory attack on a couple and their dog in Banff National Park, Alberta, Canada. The couple, whose names were not yet released, were said to be seasoned hikers who practised bear safety. They had sent a notification that they had arrived safely at camp in the Red Deer River Valley, an area of steep cliffs and rugged terrain west of Ya Ha Tinda Ranch on Friday, September 29, 2023. Later that night, park staff received a distress notification concerning a bear attack via a satellite device.

A specialist team was to be dispatched by helicopter, but bad weather grounded it, and the team went in on foot. They found a large grizzly guarding the bodies of the couple and their dog. It did not retreat and became aggressive. The bear was shot dead.

Apparently, there had been few berries due to an early frost, and the bear may have been in hyperphagia, a kind of feeding frenzy when bears try to put down huge amounts of calories before hibernation.

THE AMERICAN BLACK BEAR

The American black bear (*Ursus americanus*) is far more widespread than brown bears in North America. It is found throughout Canada and in the northwest and northeast US, with more sparse populations in many of the more Southern states. Smaller by far than the grizzly, American black bears max out at 1,100 lbs, but most are about half that weight. They are less aggressive than grizzlies, which may explain why they have not been exterminated over much of their range, as have brown bears.

Despite its name, the American black bear can also be brown, grey, white, or even bluish! There are sixteen subspecies.

Native Americans have many stories about the black bear. One legend tells of a Cherokee clan who called themselves the Ani Tsa'gu hi. Once a young boy of the clan kept disappearing into the forest, only to return to the village a little hairier each time. The elders of the tribe asked the boy what was going on, and the boy said that he had been spending time with the bears of the forest, sharing their foods and ways. He told the elders the bears had plenty of food and that the rest of the tribe could join him rather than go hungry, but first they would have to fast to prepare for the transformation.

Informing the other clans, this Ani Tsa'gu hi clan chose to follow the boy and leave the human world of struggle and hunger behind and live forever with the black bears in their forest.

Upon their departure, the Ani Tsa'gu hi informed all the other Cherokee clans, saying, "We are going where there is much food. Do not fear to kill us, for we will be ever alive."

Another Cherokee legend tells of a healing lake that black bears used. A wounded bear would swim across the lake and by the time it reached the other side it would be completely healed.

According the Gitga'at First Nation of Hartley Bay, British Columbia, the Wee'get, a raven spirit that created the world, turned every tenth black bear white to remind people of the Ice Age. These bears were called Moksgm'ol and were thought of as sacred.

The Kwakiutl people of the Pacific Northwest coast say that black bears and brown bears became enemies after a male brown bear married a female black bear, but later killed her for being lazy. In revenge, the black bear's family killed the brown bear's children. In reality, brown bears do, on rare occasions, kill black bears.

Black bears are omnivores, but dig less than brown bears and therefore eat less tubers, roots, and bulbs. They will consume grasses, buds, fruits, pine nuts, acorns, and hazelnuts. Insects and honey form part of their diet, as do fish such as salmon, trout, and catfish. They happily take carrion and will feed on the kills of other predators. Black bears also hunt live prey, such as caribou, mule deer, white-tailed deer, elk, and even moose. Birds' eggs are relished.

Though they are less aggressive than brown bears, the American black bear is the species that most humans will encounter in North America, polar and grizzly bears living in more remote areas. Predatory attacks on humans are rare, but they do occur.

The earliest account comes from August of 1784, when a Mr. Leach sent his eight-year-old son into a pasture with a horse. A black bear appeared and grabbed the boy by the throat. Leach struck it with a wooden pole and the bear fled, dragging the boy with it.

A search was mounted the next day, and the child's half-eaten body was found. The black bear rose up to defend its kill and charged at the searchers, only to be killed with three gunshots.

May 19, 1901, saw the deaths of three siblings in the jaws and claws of a black bear. Mary Porterfield, aged three, Willie Porterfield, five, and Henry Porterfield, seven, went out to gather flowers in the woods at Job, West Virginia. When they did not return, a search party led by John Weldon was sent out. Apparently, the kids had become lost in deep woods and were attacked by a bear. Their partially eaten remains were found and the bear, which was lurking close by, was shot.

July 7, 1948, the mother of three-year-old Carol Ann Pomranky saw a black bear emerge from the woods in Marquette National Forest, Michigan and grab her child. As the bear made off into the woods, she called her husband Arthur Pomranky, who was the lookout and keeper of the fire observation tower at Mission Hill.

A group of volunteers with hunting dogs was soon searching for the girl. Soon the dogs found one of the girl's shoes, and about a quarter-mile away in the woods, they found her chewed body along a stream. One man waited for the bear to return to finish its meal and then shot it.

On May 14, 1978, four boys were fishing in Algonquin Provincial Park in Ontario, Canada. Eighteen-year-old Richard Rhindress and his sixteen-year-old brother, William, invited twelve-year-old George Halfkenny and his fourteen-year-old brother Mark, and they were all at Radiant Lake.

As the afternoon wore on, the boys spread out, losing visual contact. Missing George, the others began to call out. Investigating, they found no sign of the boy. As Mark walked into the forest looking for his brother, he had no idea that the boy had been killed and dragged away by a black bear. The same animal ambushed and killed Mark. Both boys were apparently killed by a bite to the neck.

William heard the commotion and investigated. The bear circled back through the underbrush and attacked him. The killing method was the same, a bite to the neck.

Richard Rhindress then went about searching for the trio but saw no sign of them. He left and enlisted the help of the authorities, who found the bear guarding the boy's bodies for later meals.

The bear was later killed with three shots from a rifle wielded by Lorne O'Brien. The bear had covered the bodies with earth.

On August 14, 1980, two people were killed by the same black bear near the remote town of Zarma, Alberta. Both worked on an oilfield in the area. Forty-four-year-old Lee Randel Morris and twenty-four-year-old Carol Marshall. Morris had left camp between 4:45 and 5:30 p.m. and did not return for supper. Marshall went for a cigarette break with Gordon Elias about two hours later, unaware that Morris had been killed.

Suddenly the bear appeared, and the pair ran for their lives. Elias managed to climb a tree, but the bear caught and killed Marshall. Later a boy, son of the camp manager, came by and Elias shouted for him to get help. The boy returned with his father, Bud Whiting, who managed to shoot and wound the bear, causing it to retreat.

The body of Morris was recovered only after Elias was rescued. It was cached only a few metres away, covered with earth.

Rancher Sevend Satre, fifty-three, a rancher from British Columbia rode out on his horse to check on his cattle on June 14, 1996, near Tatlayoko Lake. He never returned, and his partially eaten body was found the next day, guarded by an aggressive black bear. The animal was shot, and human flesh found in its stomach. Apparently, the bear had stalked Satre whilst he was on horseback for half a mile, demonstrating it had no fear of the horse.

Bernice Evelyn Adolph, seventy-two, was a Xaxli'p First Nations elder. Black bears had been frequenting her land near Lillooet in a remote part of British Columbia. Fewer than one thousand people lived in the town. A tribal council had decided to kill the bears, but Mrs. Adolph objected.

"Our tradition is we respect our elders and the direction they give, even though it was a hard decision but that was the direction she gave us. She didn't want us to harm the bears, saying they are small, and to not hurt them," Chief Art Adolph said.

On September 1, 2011, she was mauled to death by a bear. Her remains had been found in the woods and her body had been so stripped of flesh that it took dental records to identify her.

The Suncor energy company has a mine in a remote part of Alberta. It is isolated on deposits in North Steepbank, Fort McMurray, a wild and untamed area. On May 17, 2017, Lorna Weafer, thirty-six, was walking back from the washrooms when she was attacked by a black bear. Efforts by coworkers to scare off the bear with a fire extinguisher, a water cannon, and an air horn were unsuccessful. The fatal predatory attack lasted over an hour and only came to an end when Royal Canadian Mounted Police shot the bear dead.

In September of the same year, but hundreds of miles south in West Milford, New Jersey, a black bear killed twenty-two-year-old Darsh Patel. He was one of a group of five people hiking in the Apshawa Preserve. They were warned by other hikers to leave the area due to the presence of a bear. However, they pressed on, found the animal, and began to photograph it.

The bear began to approach them, and they moved away. The hikers ran in different directions and found that Patel was missing when they regrouped. Authorities found Patel's body after searching for two hours. A black bear found in the vicinity was killed and a necropsy revealed human remains in its digestive tract.

Daniel Ward O'Connor was sleeping near a firepit at a camp in near Mackenzie, British Columbia on the night of May 10, 2015. A bear attacked the twenty-seven-year-old and dragged him from his sleeping bag. His fiancée, Jami Wallace, who slept in a nearby motorhome, discovered a trail of blood in the morning. With no mobile phone coverage, she had to drive and get O'Connor's father, Danny O'Conner.

The pair found the victim's body being guarded by a bear that was eating it. Retreating to the motorhome, Danny phoned the Mounties and a Conservation Service Officer, who shot the bear.

Two predatory black bear attacks happened within a day of each other in Alaska in 2017. On June 18, sixteen-year-old Patrick Cooper was chased and killed by a bear near Indian, Alaska. He was completing the juniors' division of the Bird Ridge trail's running race when he called his brother to say he was being followed by a bear. Searchers found his remains 1,500 feet from the trail, with a black bear lurking around them. The bear was shot in the face, but survived and ran away.

The next day, at Pogo Mine, employee Erin Jonson, twenty-seven, was killed. Jonson was doing contract work, collecting soil samples, with Ellen Trainor, thirty-eight. The bear stalked the women and attacked them from behind. Neither had time to use the bear spray they were carrying. Mine officials shot the bear.

Catherine Sweatt-Mueller's mother rang the police when her daughter did not return from checking on their dogs at Red Pine Island, Ontario. It was September 1, 2019. The investigating police found a highly aggressive black bear guarding her corpse and were forced to shoot it.

On August 20, 2020, Stephanie Blais was talking to her father via satellite phone from the village of Buffalo Narrows, Saskatchewan. The small village is remote and has only 1,014 residents. As she was in conversation, a black bear appeared from the woods and savagely attacked her. Her father, Hubert Esquirol, said he heard a gurgling sound. "I waited on the line for two minutes, and I called her name, I said hello, and there was no answer. So I was talking to her on the phone when the bear attacked her."

After waiting two minutes, Esquirol disconnected and called back. No one answered. Seven minutes later, he got a call from his daughter's husband, Curtis Blais, who had been in the cabin's kitchen about one hundred feet away. "Curtis called advising me that a bear attacked her, that he sprayed the bear with pepper spray, and the bear got more angry."

Curtis then fetched a gun and shot the bear twice, killing it, but his wife was already dead. The couple's children had been with them just minutes before the attack, but their mother had sent them inside.

The attack was unprovoked and predatory.

Laney Malavolta, thirty-nine, was walking her dogs in Durango, Colorado on April 30, 2021. She was attacked by a mother bear and two cubs that killed and ate her. Authorities killed the bears and an autopsy confirmed that they had devoured her.

At the time of writing, the latest American black bear attack took place in Groom Creek, Arizona, on June 16, 2023. Sixty-six-year-old Steven Jackson was sitting in a chair outside his campsite when a black bear grabbed him, dragged him 250 feet and began to eat him alive.

Neighbors heard his screams for help amid the struggle and tried to scare the bear away by yelling and honking horns, but to no avail. One neighbour eventually grabbed a rifle and shot the bear, killing it, but Jackson was already dead. The bear was healthy and well-fed, and the attack was classified as predatory.

The number of kills attributed to this relatively small and usually non-aggressive species may seem surprising. However, when the animal's population and range are considered, this seems less remarkable. With growing populations of black bears and humans, a small number of bears will inevitably look on humans as prey.

OTHERS

SLOTH BEARS

One of the most infamously aggressive bears is the sloth bear (*Melursus ursinus*). It is found across much of India and has a subspecies in Sri Lanka. It is a small bear with a maximum weight of 450 lbs. Its long shaggy hair gives it a distinctive "unkempt" look. Most have a cream-coloured *V*-shaped marking in the chest.

Sloth bears feed on honey, small mammals, and a wide range of fruits. Much of their diet consists of ants and termites. They have very flexible lips, and their jaws lack incisors, allowing them to suck up insects. Their tongues are long. Mother bears will regurgitate a mix of jack fruit, wood-apples, and honey to feed to their cubs.

The sloth bear's most famous feature is its long hard claws, with which it rips into rock-hard termite mounds. These claws are also formidable weapons. Being a rather small bear, they often fall prey to tigers. Sloth bears have been known to face down these giant cats with their claws.

Most attacks on humans are defensive ones, where the bears see humans as a threat. They are notoriously ill-tempered and will sometimes attack at the drop of a hat. The bears cause horrific wounds to humans with their claws, mainly targeting the head and face.

In rare cases, sloth bears will eat their human victims. One such was the sloth bear of Mysore. In the 1950s, a sloth bear began attacking, killing, and, in some cases, eating human victims in the hills around Arsikere, a city in the Indian state of Mysore. The beast used its massive claws to mutilate the faces of its victims.

British big-game hunter Kenneth Anderson became involved with the bear in a most tragic manner. He was driving from Bangalore to Shimoga to hunt a man-eating tiger when his car broke down in front of a Muslim shrine near Arsikere. The man who maintained the shrine, Alam Bux, helped Anderson to repair his car and invited him into his house for tea. Outside the house were rows of fig trees, peanuts, and a water tower with a pool. Bux had a wife, a twenty-two-year-old son, and a teenaged daughter. The family became friends with Anderson, and he would often visit, bringing them gifts, as the life of a shrine caretaker was a meagre one.

Some time later, Anderson heard stories of an aggressive bear in the area near his friend's house but was little worried, as he thought of bears as not particularly dangerous. But then he received a postcard from his old friend, urging him to come to his home.

Upon arrival, he found that Bux's son had been horrifically savaged by a sloth bear that had been eating figs outside one night. The young man later died, and Anderson promised Bux he would hunt the bear. The big-game hunter searched the forest and hills, looking into two hundred or so caves, valleys, and crevices, but found no sign of it. Believing it to have fled, he returned to Bangalore.

Soon after, he received a letter from a district forest officer telling him that the bear was back and had attacked two woodcutters, killing one. Anderson drove the long way back to hunt the bear again.

This time, he stopped in a disused house built by the Mysore Forest Department. Whilst here, a young boy came banging on his door, telling him that a bear had attacked his brother. He set out to find the victim and came across him, alive but badly savaged. His face had been ripped apart by the bear, and his stomach slashed so badly that the entrails hung out. Amazingly, he was still alive. Pushing the victim's innards in again and keeping pressure on his stomach, Anderson bravely carried the mauled man back through the jungle as night fell.

In the darkness, Anderson stumbled and broke his ankle. Refusing to leave the victim, he made a splint and sat by him with his rifle ready. By dawn, the man had died, but a forest ranger found Anderson. He spent the next two weeks in hospital, but as soon as he could, he was back out on the trail of the killer bear. It had claimed two more victims in those fourteen days.

Returning to the forest hut, he spent three days hunting the bear, until he found a row of date trees where it had been feeding. He set up a chair and a stool for his heel. He waited there with his rifle. At eleven o'clock that night, the bear appeared, eating the dates. Shining his torch on the beast, he shouted out, "Hello, bear." It reared up in rage, but before it could charge, he shot it in the chest, dropping it dead.

Anderson wrote of the beast...

Bears, as a rule, are excitable but generally harmless creatures. This particular bear carried the mark of Cain, in that he had become the wanton and deliberate murderer of several men, whom he had done to death in most terrible fashion, without provocation.

In his 1931 work *A Book of Man Eaters*, Brigadier General R. G. Burton recalls that in Chandra, central India, there was for six weeks a female sloth bear with two almost grown cubs who terrorized the jungle, killing and eating a number of people, though he does not provide details or a date. He cites an earlier book, *Wild Animals of Central India*, by A. A. Dunbar Brander. This was written in 1923, but again provides no dates and little detail.

On June 5, 2022, a sloth bear killed and partially ate two people in a forest near Panna in Madhya Pradesh. Eyewitnesses saw the bear clawing flesh from the bodies and eating it. The victims, Mukesh and his wife Gudiya Rai, had been collecting water near a temple when the bear attacked. It mauled Gudiya, killing her instantly. Mukesh tried to save his wife and was also killed by the bear. The bear then fed on the bodies for several hours.

Locals fired guns in the air to try and scare the bear away, but it didn't work. Panna North divisional forest officer Gaurav Sharma said, "We saw the bear walking from one body to another, clawing out flesh and eating it. It is a very rare occurrence. While sloth bears eat meat, they are not known to consume human flesh, not even of those they have killed."

THE ASIAN BLACK BEAR

Ursus thibetanus, known as the Asian black bear, moon bear, or Tibetan black bear, is a widespread animal. It ranges from Iran in the west, through

Pakistan, to northern India, Burma, Bangladesh, southern China, North and South Korea, the Russian far east, Taiwan, and Japan.

Seven subspecies exist. The Asian black bear can weigh up to 800 lbs, although the 800-lb one was an exceptionally large individual. Most are less than half this weight. It has shaggy black hair, but not so long as that of the sloth bear. Asian black bears have a white *V*-shaped marking on the chest, similar to the sloth bear.

Asian black bears eat almost any edible things in their environment. Fungi, roots, grasses, fruit, eggs, fish, insects, honey, fish, and carrion are all grist to its mill. Asian black bears will hunt and kill Malayan tapir, water buffalo, deer, serow, and wild boar.

A pugnacious animal, the Asian black bear, much like the sloth bear, has been known to savagely attack humans. Brigadier General R. G. Burton wrote in *A Book of Man Eaters*:

> ***The Himalayan black bear is a savage animal, sometimes attacking without provocation, and inflicting horrible wounds, attacking generally the head and face with their claws, while using their teeth also on a prostrate victim. It is not uncommon to see men who have been terribly mutilated, some having the scalp torn from the head, and many sportsmen have been killed by these bears.***

In Kashmir, India, the bears have always been a threat. At the turn of the last century, the hospital in Srinaga had many victims of black bear attacks. More recently, a study from 2000 to 2020 recorded 2,357 attacks. The conversion of wild habitat into orchards and farmlands was thought to be the reason for the attacks. The bears would kill or badly mutilate their human victims, but none were eaten.

In Japan, in 2016, four people were killed and partially eaten in Kazuno, Akita prefecture. Seventy-four-year-old Tsuwa Suzuki was picking bamboo shoots when an Asian black bear killed and ate her.

Three men, also collecting bamboo shoots, were taken by the bear in the same year.

Takeshi Komatsu, a local vet, said the same bear may have been responsible for all of the killings.

"After tasting human flesh, the bear may have learned it can eat them," he told *Kyodo News* agency.

A hunter killed a bear close to the body of Tsuwa Suzuki.

THE SUN BEAR

The smallest of all bears, the sun bear (*Helarctos malayanus*) is about the size of a large dog. They weigh up to 143 lbs. Their smooth coat is black and tan with a cream-coloured *U* shape on the chest. They have powerful jaws and long claws. Sun bears have long tongues for lapping up honey.

Sun bears are tropical creatures found in the jungles of Indonesia, Malaysia, northeast India, Bangladesh, southwest China, Burma, Thailand, Cambodia, Laos, Vietnam, and Brunei.

These little bears feed mainly on fruit, insects, and honey. They will also take birds, eggs, and small mammals. Their large claws are used for ripping open logs and tree trunks to search for insect larvae. They also use them for defence against predators. In one case, a fight between a tiger and a sun bear ended in the two animals killing each other.

Sun bears have been known to inflict hideous wounds on humans and even kill them, but they do not hunt humans for food.

THE SPECTACLED BEAR

South America's spectacled bear (*Tremarctos ornatus*) is the only ursid in the neotropics. They range through western Venezuela, Colombia, Ecuador, Peru, western Bolivia, and northwestern Argentina in a long thin, *S*-shaped distribution. They favour mountainous regions. A medium-sized bear, they reach 491 lbs. They are the closest living relatives to the giant, hyper-predatory, short-faced bears that once ranged throughout the Americas. These were the largest bears that ever lived and have been extinct since the end of the last Ice Age.

The spectacled bear is so named due to the light-coloured rings around its eyes.

Like most bears, they are omnivores, feeding on cactus, bromeliads, palm hearts, nuts, fruits, flowers, and carrion. They will kill prey as big as tapir and domestic horses and cows. Only one human death has been attributed to the spectacled bear. They do not look on man as food, and attack only when threatened.

THE CAUSES OF BEAR ATTACKS

There are two kinds of bear attack, defensive and predatory. The former is by far the most common. Bears hate to be surprised and will often attack if startled. Hunters creeping through the forest in camo or naturalists trying to make little noise run the risk of startling a bear. Bears are very defensive; if you stumble upon a bear with a kill, or get between a bear and its prey, it may attack. A female with cubs may attack if she perceives a human as a threat.

Predatory attacks may occur when the bear's natural food sources are in short supply or if the bear realizes that humans are easy to kill. Bears may begin to associate humans with food if they are fed by people or if they scavenge from rubbish dumps around human habitation. It is only a small step from this to seeing humans as food.

HOW TO AVOID BEAR ATTACK

By making noise, you warn bears that you are in the area and avoid surprising them. Various noise makers made for this very purpose can be purchased. These include air horns and bangers.

Always carry bear spray when in bear country. This is a kind of super mace that contains red pepper oil that will usually deter an aggressive bear when shot into its face.

Hiking in groups is a good idea, as larger numbers of people will often deter bears from attacking.

Keep well clear of females with cubs or any bear guarding a food source. If you are camping, store all food and rubbish in bear-proof containers and, if possible, use ropes to hang them out of reach in trees. Never take food into tents, as it may leave a scent that causes the bear to investigate.

The old adage, quoted at the beginning of this chapter—"If it's black, fight back. If it's brown, lie down. If it's white, goodnight"—used to be said as advice for bear attacks. With black bears, fight back to deter them. With brown bears, play dead and hope that they think the threat has been neutralized, and if it's a polar bear, well, there is not a lot you can do.

In his book, *Bear Attacks, Their Causes and Avoidance*, Steven Herrero suggests that the best course in all attacks is to fight, especially if there are more than one of you. Shout, make noise, hurl rocks, make yourselves look larger. Make the bear think that you are not worth expending the energy on. Predators will often back down from what they think may be a hard kill. A hunting animal in the wild can ill afford to be injured and will abandon something they think is not going to be an easy kill.

CHAPTER FIVE

HUMAN CANNIBALS

"Cannibal chiefs chew
Camembert cheese 'cause
chewing keeps 'em cheeky."

—VIVIAN STANSHALL-LABIO,
"DENTAL FRICATIVE"

On November 19, 1961, Michael Rockefeller, the twenty-three-year-old son of noted American politician, businessman, and forty-first vice president of the US Nelson Rockefeller, was travelling in a motorized boat constructed from a raft lashed to two dugout canoes. With him were Dutch anthropologist Rene Wassing and two native guides. The pair were travelling along the Aswets River in Southern New Guinea. Recently graduated from Harvard, Rockefeller had been acting as the sound recordist and still photographer. Rockefeller had become an ardent collector of native carvings. Together they were visiting remote villages of the Asmat people. The area was dangerous, and natives called it the "Lapping Death." The riverbanks were mires of mud infested by crocodiles and mosquitoes.

At the wide mouth of the river, a powerful tidal surge swamped the boat and killed the engine. The two natives swam to shore to get help, but the nearest village was a day's trek through the jungle. Rockefeller grew sick of waiting and decided to swim for the bank himself. Despite warnings from Wassing, he set out, using two empty petrol cans for buoyancy. It was the last time any Westerner saw him alive. Wassing was rescued the following day.

It took three days of the news of his disappearance to reach the states. A massive search was undertaken, involving five thousand people, but no trace of Michael Rockefeller was ever found. The Asmat were still headhunters and cannibals at the time, and it was thought that he had been killed and eaten by tribesmen.

Investigative journalist Milt Machlin investigated the case in 1969. He was told that Dutch patrols had killed several tribal leaders at Otsjanep village. Rockefeller may have been killed and eaten in revenge as he was from the "white tribe."

Paul Toohey, in his book *Rock Goes West*, says that the Rockefeller family hired a private investigator to look for their son. He was paid $250,000 to search for evidence in New Guinea. The investigator swapped a boat engine for the skulls of the three men that a tribe claimed were the only white men they had ever killed. He returned the skulls to the family, and they were convinced that one of them belonged to their son.

The eating of human flesh is the ultimate taboo, but it has been going on for a very long time. Fossil bones of an *Australopithecus boisei* unearthed in 1970 in the Turkana region of Kenya show marks from stone tools that were used to strip the flesh from the leg bone. Briana Pobiner of the Smithsonian's National Museum of Natural History said, "These cut marks look very similar to what I've seen on animal fossils that were being processed for consumption. It seems most likely that the meat from this leg was eaten and that it was eaten for nutrition as opposed to for ritual."

More recently, some 900,000 years ago, *Homo antecessor* was hunting, butchering, and eating its own kind. The bones of seven *Homo antecessor* individuals were found to have butchery marks from stone tools and tooth marks from their own species. The bones from the Spanish archaeological site Gran Dolina also displayed cracking to expose the marrow. There were many other animals to hunt, but they chose their fellow hominins. Researchers used computer models to calculate how many calories *Homo antecessor* would require per day. Then, they estimated the caloric payoffs of various animals, their own kind, and the energy that would have been needed to catch them. They speculated that *Homo antecessor* hunters would choose their prey based on a balance: the most calories for the least effort.

Cannibalism has also been recorded in Neanderthals. Eight specimens dating to 43,000 years ago were found in a cave in El Sidrón, northwest Spain. Antonio Rosas of the Museo Nacional de Ciencias Naturales in Madrid said, "There is strong evidence suggesting that these Neanderthals were eaten. That is, long bones and the skull were broken for extraction of the marrow, which is very nutritious. I would say this practice was general among Neanderthal populations."

Rosas thinks that periods of starvation caused the cannibalism.

Another group that lived between 40,500 and 45,500 years ago in Belgium shows unmistakable signs of butchery. Ninety-nine bone fragments from five individuals—four adults and a child—showed hammering to extract marrow and cut marks from carving the flesh away from the bone. Horse and reindeer bones from the cave had been processed in the same way. The victim's bones had also been used as tools to sharpen stone tools.

Unpalatable as it is, cannibalism has been part of the human condition for over a million years, and it's still with us.

I once met a man who had eaten human flesh, and he was very nice! In 2007, I led an expedition into the interior of Guyana in search of the giant anaconda. Whilst there I met a man at Point Ranch called Joseph. In 1975, he had ventured into the wild Pakaraima Mountains after a lone pilot had crashed his plane there. The man's parents had asked Joseph to find their son's body and bring it back to them. He eventually found the body, badly burned, in the wreckage of his plane. Securing the corpse in a sling made from a blanket, he began his way down the mountain. Unfortunately, on his way down he became hopelessly lost in the jungle and began to starve. After several days he was forced to eat the arm of the corpse. Quite what he told the dead man's mother and father he never revealed.

This was cannibalism through necessity, something we will not be examining in this book. There is a world of difference between eating human meat to survive, like the people stranded in the Andes after the crash of Uruguayan Air Force Flight 571 or the survivors of the sunken whaling boat the *Essex*. Here we will be looking at targeted cannibalism, through choice.

Cannibalism has made its mark on folklore from around the world as well. No legend of cannibalism is more frightening than the Native American story of the Wendigo. In the legends of the Algonquin people of the east coast Canadian forests and the plains and Great Lakes region of the US, no other creature is as feared as the wendigo. Also known as the wetiko, the windigo, the wīhtikow and several other similar names, it seems to be an avatar of winter and hunger. It is seen as a giant, humanoid figure, but skeletally thin. Its pale skin is often caked with hoarfrost. The wendigo has teeth and claws like daggers of ice and large eyes that gleam like indigo stars on its skull-like face. Sometimes it is shown with a wild mane of white or silver hair.

The wendigo is eternally hungry, and the more it eats, the more its appetite grows. It favours flesh and prizes the flesh of humans more than any other. The icy giant can fly by "walking on the winds."

But the spirit has another, more insidious power. Should a human, caught in the wilderness in winter, be forced into eating human flesh to survive, a real possibility in the wilds of Canada or the northern US, the spirit of the wendigo could possess them. Once in the thrall of the wendigo, the victim has an overwhelming craving for human flesh and becomes a cannibal. The victim then begins to change physically until they change into a small version of the wendigo itself and go on a killing rampage. Such a state is called "going wendigo" and those suspected of it were generally killed by their peers out of fear. If caught early enough, the process could be reversed by binding the victim and suspending them over a fire. It was thought that those going wendigo had green ice growing around their heart. If the green ice could be melted, the curse would be lifted.

The Seneca of western New York State warned their children not to misbehave, or Hagondes, a long-nosed cannibal clown, would steal them away in his basket.

The Greek god of time, Cronos, who had risen to power by overthrowing and castrating his own father, became so paranoid about his own children overthrowing him that he ate them all, save Zeus whom his mother Rhea replaced with a stone in swaddling cloth for the god to swallow.

Baba Yaga is a witch from Slavic folklore. She lived in a house that walked on chicken's legs and she flew about in a giant mortar, steering it with a giant pestle. The hag had iron teeth and was said to devour children.

Europe is full of stories of child-eating witches. In the Grimm's *Hansel and Gretel*, the witch keeps Hansel in a cage, fattening him up to eat.

First mentioned in the thirteenth century, Gryla is a child-eating witch (sometimes depicted as a troll) from Iceland. In early accounts, she is a parasitic beggar who wanders around asking people to give her their naughty children to eat. She lives in a cave with her lazy husband Leppaloui, the Yule Cat, a giant man-eating cat, and her thirteen goblin-like sons, the Yule lads.

The child-eating hag seems to be a universal trope. In Japan, the onibaba is a demonic mountain-dwelling hag that eats people. One story tells of a monk called Tokobo Yukei, who in the year 726 was journeying through Adachigahara. He stopped one night in a cave with an old woman. The woman told him that she needed to go out and collect firewood. In her absence, he found piles of human bones of her former victims. Upon her return, she transformed into a demonic hag and tried to eat the monk. He was saved by a statue of a bodhisattva he was carrying that came to life as he chanted sutras and slew the onibaba with magical arrows.

There is another tale that explains the origin of the onibaba. She was once a nurse to an aristocratic family. The nurse was called Iwate. The young daughter of the family had a condition that meant she could not speak, even at the age of five. A soothsayer told the nurse that the cure was to feed the girl the liver from a foetus, along with the womb of the pregnant woman. The nurse left, leaving her own child behind at the house of her employers.

Iwate took up residence in a cave and waited for a pregnant woman to come by. Years and years went by, but at last a man and his pregnant wife came by. They requested lodging inside the cave. The woman was pregnant. Just at that moment, the woman started going into labour, and the husband went out to buy some medicine. Iwate murdered the woman with a knife, cut the foetus out of her belly, and removed the liver from the child. It was only then that the nurse noticed the charm about the woman's neck. It was the one she had given to her own child. Iwate had murdered her own daughter and grandchild. She was driven mad and would attack, kill, and eat passing travellers. She finally became the onibaba.

Even in the twenty-first century, man is still eating man. Cannibalism falls into two main categories, tribal or cultural cannibalism and modern, criminal cannibalism.

CULTURAL CANNIBALISM

NORTH AMERICA

In an ancient Anasazi settlement near Mesa Verde in southwestern Colorado, excavated in 2000, scientists uncovered butchered human bones and stone cutting tools stained with human blood. A ceramic cooking pot held residues of human tissues. But the most telling of the evidence was found in human droppings. Biochemical tests revealed clear traces of digested human muscle protein in the dried coprolite, or fossilized prehistoric faeces.

Dr. Brian R. Billman, an archaeologist at the University of North Carolina, said, "Analysis of the coprolite and associated remains at last provides definitive evidence for sporadic cannibalism in the Southwest."

The biochemist who conducted the tests, Dr. Richard A. Marlar of the University of Colorado School of Medicine in Denver, said the only way this particular human protein, myoglobin, would be present in the faeces was if the person had eaten human flesh.

As mentioned above, the wendigo is the terror of the northern Indian tribes. It is a warning against the taboo of eating human meat. Yet there is a form of madness recognized by psychologists known as wendigo psychosis. Wendigo psychosis is characterized by psychiatric manifestations such as paranoia, anxiety, hallucinations, and cannibalistic urges. Typically, the onset is marked by depression, nausea, and anorexia. As the disorder advances, individuals begin to believe that they are possessed by the wendigo, leading to heightened levels of paranoia and violent hallucinations of potential victims in the form of animals to be killed and devoured. Further along, those afflicted may develop an alarming shift in perception, viewing others, even close family members, as potential prey. Once human flesh is consumed, the transformation to wendigo is considered complete and irreversible.

The pioneer and surveyor David Thompson recorded a case of wendigo psychosis in 1796 whilst living with a group of Cree between Rainy River and the Lake of the Woods. One morning, a young man woke up filled with an urge to kill and eat his own sister. Despite his family's efforts, the youth's morbid urges grew stronger. At a tribal meeting, the medicine man decided that a wendigo had possessed the boy. A sentence of death was passed on him and he was quite willing to die. He was strangled with a cord and his father burned his body to ashes.

At Smokey River in 1899, a man called Moostoos was bludgeoned with a hatchet and stabbed by members of the party he was with. He had warned them that he was about to go wendigo. He said, "I don't want to do anything to my children. It is better that they should kill me. How would it do if I should eat my little ones, and especially their noses?"

Even babies could go wendigo. Anthropologist Ruth Landes records such a case in her 1938 study *The Ojibwa Woman.*

> ***The infant son of the Shaman Great Mallard Duck was viewed by his mother's co-wife and by his half-sisters as a windigo and therefore killed. This happened during a period of starvation, when seven out of Duck's family of 16 persons died of hunger. The baby that was nursing was just crazy. He was eating his fingers up and biting the nipples off his dead mother's breasts. They knew he was to become a little windigo. His eyes were blazing and his teeth rattling. So the old woman killed the little boy.***

The best-known case of wendigo psychosis occurred in Alberta in 1879. It gained notoriety in newspapers across Canada and the US. A Plains Cree hunter called Swift Runner walked into a Catholic mission in St. Alberts. He claimed that his whole family had starved during the harsh winter. The priests became suspicious when Swift Runner began to scream in his sleep and was plagued by nightmares. He also looked well-nourished for a man whose family had all starved. He told them that he was being tormented by a wendigo.

The priests reported their suspicions to the police. The police took Swift Runner back to his camp in the forests of northeast Edmonton. Here they found the bones of his whole family, with every scrap of flesh gnawed from the bones and the marrow sucked out. Swift Runner had not eaten his family out of desperation. There was plenty of food at the Hudson Bay Company post only twenty-five miles away, no distance for a Cree hunter, even in winter. He confessed to killing and eating them all. He told Father Hippolyte Ledue, "I am the least of men and I do not merit even being called a man."

A contemporary photograph of Swift Runner shows a haunted man staring blankly as the snows swirl around him. He was hung on December 20.

A man named Napanin arrived at Trout Lake, an outpost near Wabasca, Alberta, Canada, in 1896, claiming that the wendigo was within him. He had been travelling with his wife and son when he began to see the child as a moose calf and wanted to eat him. The wife and son fled, and Napanin went looking for help for his affliction. He suffered from a constant, bone-chilling cold. Even when wrapped in six blankets, he felt as if he were freezing. Witnesses said that he made a noise like a wild bull and grew terribly swollen in both body and face. The men of the village decided that he must be killed before he began to eat the people of the village. A medicine man slew him with an axe and his body was buried with a massive pile of logs on top of it to prevent him rising from his grave. His severed head was buried separately.

CENTRAL AND SOUTH AMERICA

The Spanish recorded cannibalism among the Aztecs of Mexico whom they conquered from 1519–1521. Juan Bautista de Pomar was a mixed-race historian, descended from Aztecs. He was a researcher into Aztec history. In his 1582 book *Relacion*, he writes that after the sacrifice, the body of the victim was given to the warrior responsible for the capture. He would boil the body and cut it into pieces to be offered as gifts to important people in exchange for presents and slaves. It was rarely eaten, since they considered it of no value. However, the conquistador Bernal Diaz del Castillo recorded that some of these parts of human flesh made their way to the Tlatelolco market near Tenochtitlan. In his book written in 1568, *The Conquest of New Spain*, Diaz writes of a number of observations of cannibalism.

In the city of Cholula, he writes of seeing young men kept in wooden cages like livestock to be sacrificed and eaten. He saw scenes like this in many towns and wrote of it, "Eating human meat, just like we take cows from the butcher's shops, and they have in all towns thick wooden jail-houses, like cages, and in them they put many Indian men, women and boys to fatten, and being fattened they sacrificed and ate them."

In the Aztec capital, Tenochtitlan, he observed large pots of human flesh. The victims had been sacrificed to the flying serpent god Quetzalcoatl and the human meat was to be cooked and fed to the priests.

Another Spanish historian, Diego Munoz Camargo, echoes this in his book *History of Tlaxcala*, written in 1585. "Thus there were public butcher's shops of human flesh, as if it were of cow or sheep."

Cannibalism was confirmed in another Mexican people, the Xiximes, in a dig in 2003. Jose Luis Punzo, an archaeologist from the National Institute of Anthropology and History, said, "The newfound bones prove that cannibalism was a crucial aspect of their worldview, their identity. Through their rituals, cannibalism, and bone hoarding, they marked a clear boundary between an 'us' and 'them.' "

The Xiximes believed that ingesting the bodies and souls of their enemies and using the cleaned bones in rituals would guarantee the fertility of the grain harvest, according to historical accounts by Jesuit missionaries. The idea had long been dismissed, but the bones found in Cueva del Maguey, a hamlet built inside a huge cliffside cave, erased any doubts.

Tests showed that 80 percent of four dozen bones, found in houses dated to around 1425, bore marks and other evidence of being boiled and cut with blades of stone. For the Xiximes, the planting-and-sowing cycle was intertwined with a cycle of cannibalism and bone rituals. After each corn harvest, Xixime warriors would go out and hunt for human prey.

Most of the time the Xiximes would prey on lone men from other villages working in the fields. Other times, the Xiximes would engage small groups in forest battles. The warriors would bring the dead victims back to the village, where Xiximes would rip the bodies apart at the joints, taking care not to break the bones.

Body parts were cooked in pans until the bones emerged clean. The flesh was then cooked with beans and corn and eaten in a type of soup, part of an all-night village ritual, complete with singing and dancing, according to the old missionaries' reports.

After the feast, the bones were stored for months in treasure houses. Then, in the run-up to the annual planting season, the Xiximes would hang the bones from roofs and trees as enticements to the spirits to help the crops along.

The Jesuit priests thought of the Xiximes as the wildest barbarians in the world.

Heading farther south, Columbus himself was told, when sailing past the island of Montserrat, that every one of its native inhabitants had been killed and eaten by Carib Indians who migrated there from the northwest Amazon. He describes how a group of Arawaks, who were travelling between islands on one of the Spanish ships, reacted when they saw an island they believed to be inhabited by Caribs.

> ***The Arawaks said that this land was very extensive and that in it were people who had one eye in the forehead, and others whom they called "Cannibals." Of these last, they showed great***

fear, and when they saw that this course was being taken, they were speechless because those people ate them and because they are very warlike.

The reports of Carib cannibals ultimately led Queen Isabella of Spain to issue a 1503 decree stating that natives could be enslaved for practising cannibalism. The word "cannibal" itself is derived from Spanish *caníbal* or *caríbal*, originally used as a name for the Caribs.

Some two hundred years later in 1694, missionary Father Labat was presented with the arm of an Englishman who had been killed, butchered, and smoked by Carib Indians on Martinique. He had been one of six killed by the cannibals.

The chronicler of Cristobal Guerra, one of the first exporters of the Orinoco, Fernandez de Navarrette, recorded meeting with Caribs who had captives from another tribe whom they were transporting back to their village to kill and eat. Six of them had already been eaten.

Father Pedro de Aguado, writing at the end of the sixteenth century, explains the customs of eastern Venezuela in the provinces of the Cherigotos and Pariagotos, Carib-speaking groups, as including the eating of human flesh through both custom and habit.

It is thought that most victims of cannibalism in the region were prisoners of war from intertribal fighting. In a description of the province of Caracas, written in 1579, Fernandez de Oviedo y Valdés says, ..."Those who are killed in war or who are taken alive are eaten...and it is this interest in eating one another which is always the main cause of their wars and disputes."

He continued:

The Carib Archers, who live from Cartagena along the major part of the coast, eat human meat, and do not take slaves nor wish any who oppose them, or who are strangers to them, to live, and everyone they kill, they eat, and the women who are

killed are eaten by the women, and the boys that they want (if by chance one of these Caribs is not with the others) are eaten later. They are then taken, castrated, fattened and eaten...

The chronicler Gomara, writing in the same period, says of the Amerindians, "They eat human flesh and wage continual war against each other, the victors eating the vanquished..."

Nicholas Federmann, a conquistador of Venezuela, reported that the Ciparicotos, of Carib affiliation, "kill and devour all the prisoners that they take, both in war and by ambush."

A. Vespucio (1454–1512), a Florentine merchant, relates the following incident, concerning some Spaniards who encountered a Carib pirogue off the coast of Tierra Firme.

We were about two leagues from the coast... In the pirogue there were four young men, they were not from the same group as the others, but had been taken prisoner in another land: and they had been castrated and all were without the male member and with fresh wounds, at which we marvelled much... They said to us that they had been castrated in order to eat them and we supposed that these were the people called cannibals, very fierce, that eat human flesh...

The Spanish priest Father Joseph Gumilla, in his book *The Orinoco Defended and Illustrated*, writes...

The Caberre are a nation having many villages and people... The Carib armadas have always taken the best part of them. They are a people not only barbarous, but also cruel, whose

usual meat is human flesh of their enemies whom they seek out and persecute.

Jacinto de Caravajal, writing in the seventeenth century, said that cannibalism was part of the initiation rites of a Carib warrior, who also had to kill three *itotos* (captives of war) with his own hands. The heads, arms, and ribs of these vanquished enemies were taken back to the village as trophies. Next, following a period of six months, during which the initiate observed a restricted diet of water, manioc, and tobacco juice, a big fiesta was organized, at which a specially prepared *itoto* was killed and ritually consumed. He continued, "The ordinary food of the Caribs is cassava, banana, fish or game... They eat human flesh when they are at war and do so as a sign of victory, not as food."

In 1654, Antoine Biet, a French Catholic priest, travelled to Cayenne in South America. He wrote of the treatment of captured *itotos*. Women and children were treated well on their arrival at the victorious village, but men, after being washed, dressed, and well fed, were taken to the village longhouse. Here they were tied up and, after the women of the village had performed a dance, were tortured to death by the application of an inflammable resin to their bodies. Certain parts of the body were then removed, such as the ears, nose, and penis. These were cooked and eaten, though some of the carcass was kept so that it could be shared with allied villages.

The French Jesuit Pierre Pelleprat, also writing in this period, had a personal experience with cannibalism. He records being offered human meat as a gift.

One Arot came to me one day, very kindly, in a neighbouring village where I was preparing the baptism of a Carib cacique... wishing to make me a presentation, which I took to be a personal gift, he passed me a basket, in which there was the hand and foot of an Arawak, and then invited me to eat. I was

horrified and told him that God is angered at those who eat their enemies: the obstacle of this basket and its contents was eventually overcome and I will say no more.

Other South American tribes practised cannibalism too. John Grillet, writing in 1674, notes the occurrence in the Acoquas and Nourages, who once lived in the interior of modern French Guiana. One of the Nourage people told him "that he had come down from them (the Acoquas) four months ago, and then they had just made an end of boiling in their pots, and eating a nation which they had just destroyed."

These people waged war on the Caribs themselves and would eat them in turn. The Caribs, he notes, were reluctant to go upriver because "they are afraid of those Nourages that eat human flesh. So that when any of them go into those parts they stay there as little time as possible."

Grillet was later shown some of the Nourages' human trophies and preserved human flesh, showing that the people the Caribs used as *itotos* could easily turn the tables.

French physician and botanist Pierre Barrere recorded ritual cannibalism linked with warfare in the Caribs. This included the torture of captured men, the taking of human trophies, especially the head, which was displayed in the village longhouse, and the cannibalism of certain parts of the prisoners' bodies.

The Caete and Tupi people of coastal Brazil believed that the act of eating an enemy and the digestion of said enemy would lead to the absorption of that enemy's strength. They also believed that weakness could be absorbed that way, so would only devour a foe who had fought well. Indeed, they thought that it was an honour to be killed in this kind of ritual themselves.

One of the first Catholic bishops to arrive in Brazil, Pedro Fernandes Sardinha, was killed ant eaten by the Caete. On July 16, 1556, he and his crew were shipwrecked off Bahai and captured by the Caete, who killed them with maces and ate them.

The archipelago of Tierra del Fuego off the southern tip of South America was visited in the 1830s by Charles Darwin on his voyage aboard the *Beagle.* The Ona and Yamana people inhabited the islands. Darwin writes of them, ..."The different tribes when at war are cannibals. They are cannibals when pressed in winter by hunger, they kill and devour their old women before they kill their dogs."

The old women would sometimes run to the mountain, but they would be caught and returned to the village for slaughter. The Yamana told Darwin that dogs were more useful than old women for hunting otters. It has since been questioned if the natives were being serious in their conversation with Darwin.

The *New York Times* of July 12, 1871, printed the following story.

CANNIBALISM.

MASSACRE ON THE COAST OF PATAGONIA.

BRITISH SAILORS SLAIN AND ONE OF THE NUMBER EATEN.

The London papers print the dispatch appended, which has been received at the Admiralty:

HMS Charybdis, Sandy Point

Straits of Magellan, May 20, 1871

SIR: I do myself the honour of reporting that previous to my leaving Valparaiso in her Majesty's ship under my command, the following melancholy occurrence came to my knowledge:

The British brigantine Propontis, on her passage from Bremen to Iquique, in passing through the Straits of Magellan, touched at Port Gallant, on the Patagonian coast, on the morning of 4th of March last. On the afternoon of the same day, the master, JAMES BARNES, with three of the crew, landed

for the purpose of cutting wood. Two days elapsed, and none of the party having returned to the vessel, a second boat was sent on shore to try and discover what had become of the missing men. After a short search the lifeless body of the master was found in the bush, with a large gash across the head and both legs cut off. The men were frustrated in their attempt to bring the corpse down to the boat by the threatening appearance of a party of Indians who now approached. They therefore got back to the boat and returned to their vessel. Shortly after they got on board, a boat with a number of Indians in it was seen making for the brigantine. The cable was consequently slipped and sail made for the Chilian settlement of Sandy Point, where the vessel arrived on the 9th inst. Passing through the Straits of Magellan in this ship, I thought it right to call in at Port Gallant, and therefore anchored there on the 19th inst. I shortly afterward sent an armed party on shore in the hope of getting some intelligence as to the fate of the missing men of the Propontis, but no human being was seen. Shortly after dark the same night there were heard cries from the shore. I consequently dispatched an armed force, with instructions to the Lieutenant in charge to bring on board any Indians he might meet. On the party landing, several natives were seen, but they quickly retreated into the thick bush. After a considerable chase one man was captured. In the expectation that he might throw some light on the subject of the horrible catastrophe I took the man to Sandy Point, and handed him over to the Governor, but nothing was elicited from him. His Excellency, who is fully impressed with the gravity of the case, informed me that he had already been in communication with his Government on the matter, and that he had made arrangements to send a detachment of troops from the force under his command to the neighbourhood of Port Gallant, with the view of punishing the Indians. I fear there is but little doubt that the three men who accompanied Capt. BARNES shared his miserable fate. I have, &c."

ALGERNON LYONS, Captain.
Rear-Admiral A. FARQUHAR.

P.S.— Since writing the above I have been informed by the Governor of this settlement that the Indian (a Fuegian) whom I brought with me from Port Gallant has made a statement to the following effect: Capt. BARNES and the three men belonging to the Propontis, who had landed with him, while employed in cutting wood, were set upon by a party of Fuegians, and with the hatchets with which they were armed they slaughtered all the four Europeans. A part of the body of the captain was eaten, and the corpses of the rest of the victims thrown into the sea. The plea for the attack is that the Captain had first fired and wounded one of the Indians.

AFRICA

The oldest written reference to cannibalism seems to come from Egypt. A hymn was found in the tomb of the pharaoh Unas from 2,400 years ago. It was written in the Pyramid of Teti and runs thus:

> ***A god who lives on his fathers,***
>
> ***who feeds on his mothers...***
>
> ***Unas is the bull of heaven***
>
> ***Who rages in his heart,***
>
> ***Who lives on the being of every god,***
>
> ***Who eats their entrails***
>
> ***When they come, their bodies full of magic***
>
> ***From the Isle of Flame...***

Abu Abdullah Muhammad ibn Battutah was a fourteenth-century Moroccan traveller, explorer, and scholar. Over thirty years, his explorations took him across much of Asia, Africa, and parts of Europe and covered around 73,000 miles!

Whilst in Mali in 1350, he saw the sultan Mansa Sulayman give a slave girl to a group of visitors from a "cannibal region" who were visiting his court. The cannibals slaughtered and ate her. Then they smeared their faces and hands with her blood and thanked the sultan.

Ibn Battutah was told that the sultan always did this when he had cannibal guests, despite being a Muslim.

Emil Torday, a Hungarian anthropologist, travelled in the Congo Basin from 1907 to 1909 and recorded that the natives were surprised that Europeans did not eat human flesh.

> ***They are not ashamed of cannibalism, and openly admit that they practise it because of their liking for human flesh, with***

the primary reason for cannibalism being a "gastronomic" preference for such dishes.

Torday himself was offered part of a human thigh as a present, and other visiting Europeans were offered human meat. Apparently, man flesh was given in payment for services, and workers were disappointed if given the meat of animals. Slaves as well as captured foes were eaten like livestock.

Many healthy children had to die to provide a feast for their owners. Young slave children were at particular risk since they were in low demand for other purposes and since their flesh was widely praised as especially delicious, "just as many modern meat eaters prefer lamb over mutton and veal over beef." Such acts were not considered controversial—people did not understand why Europeans objected to the killing of slaves, while themselves killing and eating goats; they argued that both were the "property" of their owners, to be used as it pleased them.

Torday says that the natives hunted people from other tribes, especially women and children who strayed too far from their villages whilst fetching firewood or water. They were killed with poisoned arrows and spiked clubs. Their flesh was tastier than that of adult males. The meat was often smoked if not eaten on the same day. Thus, Europeans were wary of eating any smoked meat. Torday notes...

It often happens that the poor creature destined for the knife is exposed for sale in the market. He walks to and fro and epicures come to examine him. They describe the parts they

> ***prefer, one the arm, one the leg, breast, or head. The portions which are purchased are marked off with lines of coloured ochre. When the entire body is sold, the wretch is slain.***

French missionary Prosper Philippe Augouard, who settled in the Congo in the 1880s, recorded such practices as well. Some people fattened slave children to sell them for consumption; if such a child became ill and lost too much weight, their owner drowned them in the nearest river instead of wasting further food on them, he said.

Other nineteenth-century reports come from around the Ubangi River in the northern Congo. Ivory was exchanged for slaves. The ivory was exported, and the slaves eaten. According to one trader, the human meat was easier to cook and tasted better than animal meat. Sometimes as many as ten children were brought at once to be eaten.

Augouard recorded that chiefs raised and fattened herds of children like sheep or geese for slaughter and consumption. The well-to-do ate human meat rather than animals.

In other parts of the Congo, raiding parties for slaves would often eat those captured who were too young, old, or infirm to be slaves, or sold them to cannibals.

The German ethnologist Leo Viktor Frobenius wrote that "Young children caught in raids were skewered on long spears like rats and roasted over a quickly kindled large fire, while older captives were kept alive to be exploited or sold as slaves."

Likewise, British explorer Sir Samuel White Baker, in 1863, whilst speaking to a member of a Swahili-Arab raiding party, was told that their allies from the Azande tribe would eat children that they captured in raids.

> ***Their custom was to catch a child by its ankles, and to dash its head against the ground; thus killed, they opened the abdomen, extracted the stomach and intestines, and tying the two ankles to the neck they carried the body by slinging it over***

the shoulder, and thus returned to camp, where they divided it by quartering, and boiled it in a large pot.

Another man told Baker that he had once seen how several slave children were slaughtered and then served at "a great feast" held by the Azande members of a slave-raiding party.

Another German ethnologist, George August Schweinfurth, once saw a newborn baby among the ingredients assembled for a meal. He learned that the baby, whose mother was a slave, would soon be cooked together with the gourds. He also heard persistent rumours that little slave children were frequently served at the table of the Azande king. A child of six could be brought for the same price as a dwarf goat, and people preferred the taste. A living slave child was worth less than a quarter of a hippo or buffalo.

White travellers could behave in the same vile manner. James Jameson, a Scot and member of Henry Morton Stanley's last expedition, apparently paid the purchase price of a ten-year-old girl and then watched and made drawings while she was stabbed, dismembered, cooked, and eaten in front of him.

British explorer, artist, and writer Herbert Ward was witness to another such horrific case. In the late 1880s, a white trader accidentally caused the death of a slave boy whom he had rented from a Bangala headman. When he complained about the boy's unreliability, the man reacted by killing the boy with a thrust of his spear, and one day later his teenage son nonchalantly remarked, "That slave boy was very good eating—he was nice and fat."

"The pot" was the usual destination of any slave who annoyed or disappointed their owner, and "light repasts off the limbs of some unfortunate slave, slain for refractory behaviour," were served in the area on a fairly regular basis.

French explorer Maurice Musy, whilst travelling along the Ubangi River, declined an opportunity to purchase a young slave girl despite there being human flesh cooking in pots, and he was aware that the locals were cannibals and child meat was much in demand. In a letter to his friend Albert Dolisie, he wrote, "Maybe at this hour she is being eaten. That is very likely." Oddly, he later bought a young slave boy who was destined for the pot.

During the 1892 to 1894 war between the Congo Free State and the city-states of Nyangwe and Kasongo, in the eastern Congo, the Batetela people, led by Gonfo Lutete, who were allied to Belgian commander Francis Ernest Joseph Marie Dhanis, ate the corpses of their foes killed in the war. Medical officer Captain Sidney Langford Hinde noticed the bodies of the dead and wounded going missing. Later he saw the Batetela dropping human heads, arms, and legs on the road.

Following the defeat of Nyangwe, Hinde records days of cannibal feasting "during which hundreds were eaten, with only their heads being kept as mementos. Thousands of men smoking human hands and human chops on their campfires, enough to feed his army for many days."

The Myangwe people rioted in a rebellion, but it was swiftly put down and a thousand more of them killed. The cannibals cleaned up the mess. Another Belgian officer wrote a letter home that said:

> ***Happily Gongo's men...ate them up (in a few hours). It's horrible but exceedingly useful and hygienic... I should have been horrified at the idea in Europe! but it seems quite natural to me here. Don't show this letter to anyone indiscreet!***

Hinde himself commented, "The cannibals disposed of all the dead, leaving nothing even for the jackals, and thus saving us, no doubt, from many an epidemic."

The Zappo Zap people of the Kasi region of the Congo had been cannibals from time immemorial. They did not just eat their enemies in war, but favoured human meat over all others. They kept, raised, and traded slaves as food. The local value of a slave was less than that of a pig. "If there is as much to eat on a man as on three goats, he brings the price of three goats," a settler told the missionary Samuel Lapsley. Human meat was roasted or fried like bacon.

When Hermann Wissmann, the German explorer, met the Zappo Zap in 1883, he gave them a gift of brass wire and cloth. In return he was given a number of slaves whom he later released and educated.

Fierce warriors who fought with steel spears and poisoned arrows, the Zappo Zap were the main allies of King Leopold's forces. When villages refused to pay extortionate taxes of rubber, goats, food, and slaves, the Zappo Zap massacred fourteen communities. Missionary William Henry Sheppard was sent from the Southern Presbyterian mission to investigate. He saw over forty corpses, mostly defleshed. He was told that the meat had been eaten and that other bodies had been processed and the bones discarded.

Roger Casement, a British diplomat, wrote in 1903 from Lake Tumba in the Congo to a consular colleague.

"The people round here are all cannibals... There are also dwarfs (called Batwas) in the forest who are even worse cannibals than the taller human environment. They eat man flesh raw! It's a fact."

He added that assailants would "bring down a dwarf on the way home, for the marital cooking pot... The Dwarfs, as I say, dispense with cooking pots and eat and drink their human prey fresh cut on the battlefield while the blood is still warm and running. These are not fairy tales...but actual gruesome reality in the heart of this poor, benighted savage land."

Laszlo Magyar, a Hungarian explorer, lived for seventeen years in Angola and married the daughter of the king of Bie province. He records repeatedly seeing the consumption of young slaves and victims of slave raids from other parts of Central and Western Africa. The baked bodies of deliberately fattened slave children were served as delicacies, as young children's flesh was said to be the best food of all.

Englishman George Gillman Rushby (1900–1968) was an interesting fellow. Through the years, Rushby had been a poacher, gold digger, guano prospector, ivory hunter, bouncer, and finally a game warden. We will be meeting him in volume two of this work, when he bravely tackles the man-eating lions of Njombe.

When ivory hunting in the Congo, he had brought down two large tuskers, and the bodies of the elephants were being carved up for meat by the local pygmies. Suddenly a heavily armed group of a dozen men arrived and asked if they could have some of the meat. Rushby agreed, and later their leader chatted to him. He told Rushby that they did not like the French, but not to worry; even in cannibal country, he was known as an Englishman. When Rushby asked if cannibals made such distinctions, the leader told him that he himself was a cannibal. Rushby asked if the man preferred animal or human meat, and the cannibal told him that human meat is nicely marbled, rather like prime beef, variegated with streaks of fat. He also commented that, being lighter than human meat, it bobs around in the pot as if it were alive. He much preferred human meat.

Rushby asked, in jest, for some human meat since they had eaten his elephants. The cannibal agreed, saying he had much human meat drying on the racks at his camp.

Next day, wishing to see a genuine cannibal camp, Rushby dropped by on his way to another hunting ground. There were numerous racks with meat drying over fires. Rushby examined it and found that it was not gorilla or monkey, but human. The cannibal leader gave him some before he went on his way.

Rushby buried the meat later, unseen by the other men. His actions were to have terrible results. Rushby found later that one of the porters had brought his wife along. This was strictly prohibited, as a woman in the camp, according to Rushby would cause fights, unrest, and even murders as men fought over her. He was going to fire the man, but he was so contrite that Rushby said he would reemploy him after he took his wife home, a two-week round journey, but would dock him the fortnight's wages.

After three days, the man reappeared, and Rushby was confused that he was back so quickly. The headman of the porters explained that the woman had not been taken home, but taken out of the camp and killed. They had all been eating her flesh for the last three days! The men had seen the hunter take a cut of human meat from the cannibal's camp and presumed that he had eaten it and was one of them, a cannibal himself. Horrified, Rushby ended his hunting trip and left the Congo for East Africa.

American writer and adventurer William B. Seabrook (1884–1945) travelled to West Africa in the 1920s to try cannibalism himself. The natives did not allow him to join in their cannibal rituals. Seabrook eventually got some cuts of human meat from a Parisian hospital, which he cooked and ate himself. His verdict:

It was like good, fully-developed veal, not young, but not yet beef. It was very definitely like that, and it was not like any other meat I had ever tasted. It was so nearly like good, fully developed veal that I think no person with a palate of ordinary, normal sensitiveness could distinguish it from veal. It was mild, good meat with no other sharply defined or highly characteristic taste such as for instance, goat, high game, and pork have. The steak was slightly tougher than prime veal, a little stringy, but not too tough or stringy to be agreeably edible. The roast, from which I cut and ate a central slice, was tender, and in colour, texture, smell as well as taste, strengthened my certainty that of all the meats we habitually know, veal is the one meat to which this meat is accurately comparable with.

Apparently, Seabrook too his meat with rice and a bottle of wine.

ASIA

The Official Chinese Dynastic Histories are a series of works chronicling the history of China from 91 BC to 1739. It contains more than three hundred references to cannibalism. Most of these are the results of famine or war. Baengt Patterson, in his article *Cannibalism in the Dynastic Histories*, published in the *Bulletin of the Museum of Far Eastern Antiquities*, notes that, apart from war and famine, revenge and medical reasons were chief among the reasons for the consumption of human flesh. Individuals and state officials engaged in cannibalism for revenge. The liver and heart in particular were eaten. This was common from the fourth to the tenth centuries, but occurred as recently as 1912.

In war, it was permissible to eat the enemy not just for lack or provisions, but for revenge. The concubines and servants of the foe could be eaten too.

In the sixth century, Emperor Wu of Liang, in southern China, allowed prisoners of war to be traded for food. They were caged and whenever there was a demand for meat, some of them were taken out, cut, broiled, and consumed. Later, when the usurper Hou Jing was defeated, parts of his body were said to have been salted and distributed in the regions that had suffered the most from his wars, where they were boiled in stews and consumed by angry mobs.

At the end of the Sui Dynasty (581–618), a wealthy man named Zhuge Ang threw a great ball attended by hundreds of guests. A pair of teenage twin brothers were boiled together with pigs and sheep. On the basis of this and similar accounts, the Chinese author Zheng Yi concluded that "the rich competed in wealth, a sport that included competition over cannibalism, as one way of surprising one's guests with an exotic novelty food."

From 618 to 619, the newly established Tang Dynasty was struggling to control the whole country. Rebels had attempted to conquer a district near Louyang, causing a siege. Cannibalism occurred on both sides. Rebel soldiers kidnapped children, then steamed and ate them. Their leader said: "Of all the delicious things to eat, none surpasses human flesh. As long as there are people in neighbouring districts, we have nothing to fear from famine." He is said to have used an upturned bell with a capacity of two hundred bushels to cook the flesh of women and children, which was then distributed among 200,000 soldiers.

The *Bencao Shiyi*, a book of medicine complied by Su Jing in the eighth century, records that eating human flesh was a highly effective medical treatment. In love and respect for their elders, children were recorded to slice of parts of their flesh to feed to ill elders. The *Official Chinese Dynastic Histories* record more than 110 cases of such voluntary offerings that took place between the early seventh and the early twentieth century.

> ***The devoted daughter-in-law would tie her thigh or her arm very tightly with a piece of clothing. She would then use a very sharp knife to quickly slice off a piece from her upper arm or upper thigh. The flesh would immediately be mixed in with soup or gruel, which had been heated in preparation, and this would then be offered to the dying mother-in-law or father-in-law.***

Children were also killed and their flesh eaten for medical reasons.

In the thirteenth century, after returning from China, the great explorer Marco Polo wrote:

> ***In the kingdom of Fuzhou in the south-east of the country that they eat all manner of foul things and any kind of meat, including human flesh, which they devour with great relish. They will not touch someone who has died of natural causes, but if he has been stabbed to death or otherwise killed they eat***

him all up and consider it a great delicacy. Soldiers regularly drank the blood and ate the flesh of those they had killed.

The *Benco Gangmu*, another Chinese medical book, written in the late sixteenth century, lists thirty-five parts of the human body considered to be cures for various diseases, including tuberculosis.

The transition from the Ming to Quing dynasties in the seventeenth century was a time of decades-long conflict. Frequent famines led to cannibalism, sometimes on a large scale. During a famine in 1622, government troops took the providing of human flesh into their own hands, openly butchering and selling people in a market where 600 grams of flesh could be exchanged for 40 grams of silver.

In 1640, a drought in Henan and Shandong became so bad that women and babies were arrayed in the market as human food and were sold by the slaughterers just like mutton and pork. Sometimes women and children were slaughtered in the back rooms of butcher shops while customers were waiting for fresh meat.

A few years later in Sichuan, hundreds of the young and weak were kidnapped, killed, and eaten; in the markets, men's flesh was sold at a somewhat lower price than that of women, which was considered tastier.

In Shaanxi Province in the early Quing period, cannibalism became so common that local government sanctioned the sale and eating of human flesh. Butchers legally killed and rendered humans and sold their meat. Restaurants had dishes created from human flesh.

Cannibalism in China was not restricted to the distant past. During the Taiping Civil War, from 1850 to 1864, humans were kidnapped, butchered, and sold at market. Human hearts were very popular. Zeng Guofan, the general who repressed the rebellion, confirmed the open sale of human flesh in his diary, once even complaining about its high price.

Up to the end of imperial China, cannibalism was still being recorded. During the great famine of Northern China, 1876 to 1879, humans were on the menu, with eyewitnesses reporting the sale of human flesh in markets and butcher shops.

James Wheeler Davidson, a Canadian journalist, said in his 1903 book *The Island of Formosa: Past and Present* that islanders who rebelled against Chinese rule were killed and sometimes eaten by the Chinese army.

> ***One horrible feature of the campaign against the savages was the sale by the Chinese in open market of savage flesh... After killing a savage, the head was commonly severed from the body and exhibited... The body was then either divided among its captors and eaten, or sold to wealthy Chinese and even to high officials, who disposed of it in a like manner. The kidney, liver, heart, and soles of the feet were considered the most desirable portions, and were ordinarily cut up into very small pieces, boiled, and eaten somewhat in the form of soup. The flesh and bones were boiled, and the former [latter?] made into a sort of jelly... During the outbreak of 1891, savage flesh was brought in—in baskets—the same as pork, and sold like pork in the open markets of Tokoham before the eyes of all, foreigners included. Some of the flesh was even sent to Amoy (on the mainland) to be placed on sale there. It was frequently on sale in the small Chinese villages near the border, and often before the very eyes of peaceful groups of savages who happened to be at the place.***

During the Cultural Revolution (1966–1976), three million people were murdered. Hundreds of people were killed and eaten due to been thought of as "class enemies." Chinese writer Zhen Yi, who lived through the events, recorded many eyewitness accounts of such events. In one case in Mengshan County, a schoolteacher heard that consuming a young girl's heart could cure disease. He publicly denounced one of his teenagers as a member of the enemy faction, causing a mob to killer her. Afterwards he cut out and ate her heart.

Another teacher from Cangwu County told Yi that the deputy at his school was killed and eaten by pupils and other teachers. He confessed to eating some of the man's flesh.

In another case from Wuxuan County, three brothers were beaten to death as supposed enemies, then afterwards their livers were cut out, baked, and consumed.

An account recorded in the official Cultural Revolution annals of Wuxuan County, Guangxi, makes for horrific reading.

> ***On July 10, 1968, a criticism rally was held in front of the Shangjiang Town hall, Sanli district. During the ensuing chaos, Liao Tianlong, Liao Jinfu, Zhong Zhenquan, and Zhong Shaoting were beaten to death. Their bodies were stripped of flesh, which was taken back to the front of the brigade office to be boiled in two big pots. Twenty or thirty people participated in the cannibalism. Right out in the open, they boiled human flesh in front of the local government offices.***
>
> ***He made two trips to Guangxi, spoke to family members of victims of cannibalism and even to some of the cannibals themselves who appeared to have gotten off with light sentences.***

Song Yongyi of California State University says:

> ***It was state sponsored cannibalism. It is not that Chairman Mao himself instructed local Communist Party offices to do these kinds of things. But the head of the Revolutionary***

Committee and directors of the local militia headquarters organized militia men to engage in this kind of animal behaviour. And they represented the state and the traditional party and government establishment.

In total, 421 cases of cannibalism are recorded in thirty-one of the seventy-five counties of Guangxi Autonomous Region. Cannibalism took place at county fairs, according to Yongyi.

Thousands of people participated in the fairs.

A fair was used as an occasion to organize struggle sessions against so-called class enemies, landlords, rich peasants, former members of the Nationalist KMT party, but also communists who did not agree with Mao's policies. And they just kill those victims and they cut their chests, they prise the hearts and livers out and just eat them. At least 10,000 people participated.

Song Yongyi found a complete set of secret archives covering the time of the Cultural Revolution in Guangxi. The documents were produced by a Chinese Communist Party work team that was sent to the Autonomous Region to investigate excesses of the Cultural Revolution, once it was over, so perpetrators could be brought to justice. The texts ran to thirty-six volumes!

Yongyi was horrified by whet he read.

When first I read those kinds of things, I felt so surprised. Unbelievable. The second stage, when I read all of those cases I feel my mind is just frozen. I lost all feeling. You can't believe it, there are too many!

He continues:

The theory is that the cannibalism was inspired by hatred, resulting in class struggle. But the hidden motivation is not so revolutionary. It was their personal desire. Because those people believe that when they eat other people's livers, other people's hearts, it will help them to have a long life.

Let people know the truth. Let people know the horrible consequences of the Cultural Revolution. And take history as a mirror to get a lesson from history. Let all Chinese people know that we should prevent this kind of tragedy from happening again.

In the mid-1990s, journalists from Hong Kong uncovered a horrid underground market in human foetuses in both Hong Kong and mainland China. Aborted foetuses from hospitals were being sold for over two hundred dollars each for human consumption. Eating the foetuses was said to cure asthma and rejuvenate whoever ate them.

Film director Fruit Chan Gor was repeatedly fed a medicinal soup by his doctor after an accident. He later found out the broth was made from human foetuses.

In May 2012, South Korean customs confiscated more than 17,000 "health" capsules smuggled from China that were found to contain human flesh, most likely extracted from aborted foetuses or stillborn babies.

Almost a year prior to this confiscation, one of the most authoritative Korean TV channels, Seoul Broadcasting System (SBS), produced and broadcast a documentary on these smuggled health capsules. The SBS producers first asked around in traditional herbal stores in South Korea about rejuvenating foetus pills. Some stores fearfully responded there are no such pills. Others believed that those pills are made of placentas. The team finally found someone who revealed that those pills are ordered on request. The team then followed this Korean dealer and traced the pills to their origin, a hospital in northeastern China and an apartment that processed, by drying and powdering, aborted or stillborn babies in a rural area. To confirm the contents of the pills, the SBS documentary team took them to two DNA test centres, one of which was South Korea's National Forensic Service (NFS). The results of tests showed that the pills' contents were 99.9 percent identical with human DNA. The DNA test result at the NFS showed human male DNA, therefore it could be possibly made of human placentas or male foetuses. It is unclear if the pills were made of placentas or foetuses, and whether or not the consumers knew what the pills were made of. The tremendous market price difference between placenta capsules and these health capsules, however, revealed a suspicious difference between them.

During famines in North Korea, cannibalism emerged. In 1998, several refugees reported that children in their neighbourhood had fallen victim to this custom. One said that "his Neighbors (had eaten) their daughter," another knew a woman who had eaten her two-year-old child, and a third said that "her neighbour killed, salted and ate an uncared-for orphan."

British journalist Jasper Becker interviewed refugees from North Korea and was told of a number of instances of cannibalism. A former soldier told him that starving parents "kill and eat their own infants...in many places." Other parents sold or abandoned their children, and the fate of parentless children was often to be killed and eaten. Becker heard of a couple executed for murdering fifty children and selling their flesh, mixed with pork, in the market. Another executed woman was accused of having killed eighteen children for food. In another case, a whole family of five was executed for luring small children into their house, drugging them, and chopping up their bodies for consumption.

In 2009, a man was executed in Heysan for killing a girl and eating her after supplies to the city dwindled due to unsuccessful government attempts at currency reform. A famine in northern provinces caused another wave of cannibalism in North Korea. One informant said: "In my village in May, a man who killed his own two children and tried to eat them was executed by a firing squad." Others stated that "some men boiled their children before eating them." The cases were confirmed by an official of the ruling Workers Party.

The Mongols were said to eat human flesh to instill fear in their foes, according to Matthew Paris, a thirteenth-century Benedictine monk. When they invaded Europe and reached the Austrian town of Wiener Neustadt, their soldiers ate old and deformed women right away, while virgin girls and beautiful women were gang-raped to death and then eaten; their breasts were cut off and served to the Mongol leaders as special delicacies.

During the Mongol-led Yuan Dynasty (1271–368), a contemporary Chinese writer recorded that Mongol soldiers did not hesitate to sacrifice civilians for their culinary pleasure: "Young children were the most appreciated; women came next and men last." He also criticized their cruelty, stating that victims were roasted alive (on iron grates) or boiled alive (by placing them "inside a double bag...which was put into a large pot"). His account indicates that breasts were particularly prized. If there were more corpses around than needed, they were sometimes the only part of a woman's body that was eaten.

The Italian missionary explorer Odoric of Pordenone (1280–1331) journeyed through India, Sumatra, Java, and China. He recorded that the inhabitants of Lamuri, a kingdom in northern Sumatra, purchased children from foreign merchants to slaughter them and eat them. Odoric states that the kingdom was wealthy and there was no lack of other food, suggesting that the custom was driven by a preference for human flesh rather than by hunger.

The German botanist and geologist Friedrich Franz Wilhelm Junghuhn (1809–1864) was fed some soup by a Batak raja in Sumatra. The broth was made with human flesh. The host was genuinely surprised to learn that Europeans did not like to eat human flesh, which in Sumatra was widely praised as particularly tasty. At that time, captured enemies and convicted criminals were generally eaten, and some wealthy men bought slaves for fattening and consumption.

On the neighbouring island of Borneo, the native Dayaks were said to eat the meat of their victims on head-hunting expeditions. The first white raja of Sarawak, James Brooke (1803–1868), collected eyewitness accounts of the consumption of killed enemies after war campaigns. He also heard that in some areas, a fat child was traditionally served at the harvest festival of Makantau.

Carl Alfred Bock (1849–1932) was a Norwegian explorer who visited Borneo in the 1870s. Brock met a Dayak chief named Sibau Mobang, who told him that "his people did not eat human meat every day," but rather in the context of "head-hunting expeditions." Mobang had just returned from such an expedition, in which "no less than seventy victims, men, women and children," had been killed and partially eaten. Bock also met a local priestess who said that human "palms were considered the best eating," together with "the brains, and the flesh on the knees." These parts were always eaten, even if the rest of the body was not.

American naturalist and founder of the American Museum of Natural History Albert Smith Bickmore visited Borneo in the 1860s. He discovered that the Dayaks ate human meat from captured enemies and those found guilty of crimes. They found human meat uniquely tasty.

In 1999, ethnic conflict exploded into violence in western Borneo. The native Dayaks resented the Madurese who had move to Borneo from Indonesia and accused them of theft. More than five hundred people, including babies, were killed and cannibalised in the unrest.

On the road between the town of Singkawang and the village of Montrado, five severed heads were displayed at checkpoints along the way, including those of a teenage boy and a middle-aged couple. Young warriors, armed with guns and wooden spears and smeared with blood, walked along the road openly carrying the hearts and livers of their victims as women and children looked on.

Reporter Richard Lloyd Parry witnessed the horror unfolding and saw Dayaks roasting a human body by the road. A young Dayak man boasted that he had taken part in four killings of Indonesian settlers from the island of Madura. "We caught one of them this afternoon and we killed it and we ate it, because we hate the Madurese."

Some Dayaks tried to prevent the killings. Elias Ubek, a Dayak chief of the village of Montrado, said that at least seventy Madurese had been killed and beheaded in his village alone. He said he had seen six or seven children with their heads cut off. "Some are shot first, some are stabbed to death. They don't care about women, children; they kill everyone, including babies. They chop their heads off and they eat them."

Mr. Ubek was threatened with death by his own villagers after giving shelter to two families who had been tied up and were about to be killed by Dayak warriors. "The people trying to kill them had come from another district and they were so angry, I was almost killed myself. I am their leader and I cannot cool them down."

Mr. Ubek's eight refugees boarded a military convoy which was passing through the area attempting to save Madurese fugitives. At least 150 soldiers in a dozen trucks and two armoured cars were outnumbered by a mob of Dayak warriors who followed them down the jungle road. The Dayaks attacked and the Indonesian security forces opened fire, killing five tribesmen.

The military stopped the Dayak advance outside West Kalimantan's regional capital, Pontianak, and some two hundred Dayaks were killed when they tried to break through army lines.

The Dayaks looked on the Madurese as animals after they broke tribal laws involving land, and then thought it fine to hunt, kill, and eat them.

In 2001, the tensions exploded into violence again. Dayak warriors, wielding spears and long knives, attacked a group of Madurese near the village of Parangian. The victims included pregnant women, children, and old men. 118 were killed in the attack.

Bahang Djimat, secretary of the Dayak Community Organization of Central Kalimantan, said, "The problem is, the Dayaks are very angry. They are uncontrollable. They don't recognize whether they are women or children. They just see them as animals that have to be destroyed."

More than 60,000 Madurese fled Borneo on rescue ships. The Dayaks boasted that they had killed 2,000 Madurese, in many cases cutting off their heads as trophies, drinking their blood, and cutting out their hearts and eating them on the spot.

The Indonesian government reported that 428 people had been killed, but many more were unidentifiable due to the heads being taken from the corpses.

After police promised to take the refugees to safety, hundreds emerged from hiding. As they waited for transportation to a refugee camp, a group of armed Dayaks appeared.

The Dayaks killed a woman and her child, witnesses told the *Associated Press*, and kicked the woman's head down the street. They then forced 118 Madurese into trucks, took them to a nearby football field, and butchered them.

A joint police and military force in Sampit raided the Hotel Rama, which had served as a command centre for the headhunters. Police arrested about seventy-five Dayak warriors at the riverfront hotel and seized hundreds of spears, blowguns, and machete-like knives called *mandau*. The fighters were stripped to their underwear and carted off in trucks.

Twenty-five more of them were arrested near the hotel and two were shot dead. Behind the hotel, five severed heads were found floating in the river near the hotel dock. A hundred heads were found hidden under the hotel. The Dayaks had planned to distribute them among their villages as trophies.

AUSTRALASIA

In the colonial period of Australia, many explorers, anthropologists, and natural historians recorded cannibalism among the Aborigines.

American anthropologist Alfred William Howitt (1830–1908) explored Australia in the nineteenth century. As well as studying the natural history, geology, and peoples of Australia, he led the rescue expedition in search of the lost Burke and Wills expedition in 1861 and found the one survivor, John King. He recorded his study of the Aborigines in his 1904 book *The Native Tribes of South East Australia.* He writes that the killing and eating of small children among nomadic tribes was common. It was used as a form of population control and because mothers had trouble carrying two young children not yet able to walk.

Geza Roheim (1891–1953), the Hungarian anthropologist, whilst in Australia in the 1920s, was told by the Aborigines, "Years ago it had been custom for every second child to be eaten—the baby was roasted and consumed not only by the mother, but also by the older siblings, who benefited from this meat during times of food scarcity." One woman told him that her little sister had been roasted but denied having eaten of her. Another "admitted having killed and eaten her small daughter," and several other people he talked to remembered having "eaten one of their brothers."

Roheim found that there were two reasons for the infant cannibalism.

> ***When the Yumu, Pindupi, Ngali, or Nambutji were hungry, they ate small children with neither ceremonial nor animistic motives. Among the southern tribes, the Matuntara, Mularatara, or Pitjentara, every second child was eaten in the belief that the strength of the first child would be doubled by such a procedure.***

As a rule, only babies that had not yet been named, a process that usually happened at age one, were killed and eaten. Yet in times of hardship, children as old as four would be consumed. Babies were killed by their mother, while a bigger child "would be killed by the father by being beaten on the head." A parish priest in Broome, western Australia, recorded that a four-year-old was killed and eaten by its mother in 1904.

Self-taught Irish anthropologist Daisy May Bates (1859–1951) conducted fieldwork amongst several Indigenous nations in western and southern Australia. She knew an Aboriginal woman who one day left her village to give birth a mile away, taking only her daughter with her. She then killed and ate the baby, sharing the food with the little daughter. After her return, Bates found the place and saw the ashes of a fire with the baby's broken skull, and one or two charred bones in them. She stated that cannibalism was rife in the central western and central tribes.

She records that members of other tribes were hunted like animals and that people shared human meat like kangaroo or emu meat. The bodies were roasted in deep holes.

> ***The older men ate the soft and virile parts, and the brain; swift runners were given the thighs; hands, arms or shoulders went to the best spear-throwers, and so on.***
>
> ***Cannibalism had been rife for centuries in these regions and for a thousand miles north and east of them. Human flesh was not eaten for spiritual reasons and not only due to hunger; rather it was considered a favourite food.***

Carl Sofus Lumholtz (1851–1922), a Norwegian explorer and ethnographer, spent ten months with the Aborigines of North Queensland. He noted that infants were commonly killed and eaten, especially in times of food scarcity. He notes that people spoke of such acts as an everyday occurrence, and not at all as anything remarkable. He notes that:

> ***The greatest delicacy known to the Australian native is human flesh. The appetite for human flesh was the primary motive for killing. Unrelated individuals and isolated families were attacked just to be eaten and any stranger was at risk of being pursued like a wild beast and slain and eaten. Acquiring human***

> ***flesh is this manner was something to be proud of, not a reason for shame.***

He stresses that such flesh was nevertheless by no means a daily food, since opportunities to capture victims were relatively rare.

English explorer, painter, and naturalist George French Angas (1822–1886) spent seven years in Australia and became director of the Australian Museum in Sydney. He records the kidnapping, killing, and eating of several children near Lake Alexandrina in southern Australia.

The huge and still ill-explored island of New Guinea has been home to cannibal cultures.

James Chalmers (1841–1901) was a missionary from the Western Highlands. He spent twenty-three years visiting villages along the coast of New Guinea, converting natives to Christianity. Chalmers and his missionary colleague Oliver Tomkins waded ashore at the ironically named Risk Point on Goaribari Island, New Guinea, on Easter Sunday 1901.

The natives warmly greeted the two missionaries and invited them back to their longhouse for refreshments. Then, without warning, the natives attacked, killed, and dismembered their two visitors, passing the limbs to the women to be cooked, mixed with herbs.

Many years later, in the mountains of New Guinea, a strange disease arose from the eating of human flesh. The hill tribes were discovered by Western explorers in the 1930s. In the '50s, when researchers first visited them, they heard of a weird malady. Among a tribe of about 11,000 people called the Fore, up to two hundred people a year had been dying of an inexplicable illness. They called the disease *kuru*, which means "shivering" or "trembling."

Kuru caused a loss of control of the limbs, followed by a loss of control over emotions. Victims would laugh hysterically as if poisoned by the Joker. Loss of bodily functions would follow, and within a year the sufferers could not even stand. The disease was always fatal.

Many tribespeople thought it was the result of witchcraft. The disease primarily hit adult women and children younger than eight. In some villages, there were almost no young women left.

In 1961, Shirley Lindenbaum, a medical anthropologist with the City University of New York, visited the area. A long list of contaminants had been ruled out, and most thought the condition was genetic. However, Lindenbaum knew it couldn't be genetic, because it affected women and children in the same social groups, but not in the same genetic groups. She also knew that it had started in villages in the north around the turn of the century, and then moved south over the decades.

Lindenbaum found that the Fore people ate their own dead.

If the body was buried it was eaten by worms; if it was placed on a platform it was eaten by maggots; the Fore believed it was much better that the body was eaten by people who loved the deceased than by worms and insects.

The brains were removed by women, mixed with edible ferns, and cooked in tubes of bamboo. The rest of the meat, except the gall bladder, was roasted on fires. Women performed this as it was thought that their bodies could house and tame evil spirits.

So, the women took on the role of consuming the dead body and giving it a safe place inside their own body—taming it, for a period of time, during this dangerous period of mortuary ceremonies.

But women would occasionally pass pieces of the feast to children. They ate what their mothers gave them, until the boys hit a certain age and went off to live with the men. Then, they were told not to touch that stuff.

It seemed that the eating of brains and kuru were linked. The US National Institutes of Health injected infected human brain into chimpanzees and watched symptoms of kuru develop in the animals months later. The group won the Nobel Prize for the research.

At first it was thought to be a virus. But it was no virus or bacterium. It was an entirely new infectious agent, one that had no genetic material, could survive being boiled, and wasn't even alive in the traditional sense.

Stanley B. Prusiner, an America neurologist and biochemist, pinned it down. The infectious agent was a twisted protein capable of causing normal proteins on the surface of nerve cells in the brain to contort just like them. They were dubbed "prions" and would eventually misfold enough proteins to kill pockets of nerve cells in the brain, leaving the cerebellum riddled with holes, like a sponge.

It is thought that kuru started when somebody in the population developed Creutzfeldt-Jakob disease, a rapidly progressive, invariably fatal neurodegenerative disorder believed to be caused by an abnormal isoform of a cellular glycoprotein known as a prion.

About one in a million people develop Creutzfeldt-Jakob disease, and usually the infected tissue does not come into contact with other people. But with the Fore, the prions were passed on from person to person, sometimes taking years to show their effect.

The custom of eating the dead was stopped. Michael Alpers, a medical researcher at Curtin University in Australia, tracked kuru and found that the last death from it was in 2009. By 2012, the epidemic was declared over.

The Korowai, of southeastern New Guinea, were unaware of any other people than themselves until the 1970s. They were first contacted in 1974 by a group led by Peter W. Van Arsdale. The Korowai have been reported to practice ritual cannibalism up to the present day as a form of criminal punishment.

In May 2006, tour guide Paul Raffaele led a group from the Australian news program *60 Minutes* deep into the jungle to meet the Korowai. After filming for some days, the crew were allegedly approached by a man who claimed his six-year-old nephew Wa-Wa had been accused of being possessed by an evil spirit or *khakhua*. After the death of his mother and father, the clan suspected that he had killed them using black magic. The boy was due to be killed and eaten. The *60 Minutes* crew refused to intervene, but Raffaele tried to save him by sending a team from the current affairs show *Today Tonight* to rescue the boy. However, the group were deported over visa issues before they could reach the boy. His fate remains unknown.

Raffaele said he met the tribe via his Sumatran guide Kornelius, who visited the Korowai around 2004.

> ***They put a test to him, to determine whether they'd allow him to stay or not. One night they gave him a pack of meat and told him it was human. If he ate it, he could stay with them and if he didn't, then they'd tell him to leave. He ate it and so he became very close to them.***

When first going upriver to meet the Korowai, Raffaele and Kornelius were attacked by the tribe, who thought they had angered a river god. They shot arrows until Kornelius finally made peace with them. Talking of the cannibalism, Raffaele said:

> ***For the Korowai, if someone falls out of a tree house or is killed in battle then the reason for their death is pretty obvious. But they don't understand microbes and germs (which the rain forests are rife with) so when somebody dies mysteriously to them (of a disease), they believe it is due to a khakhua, a witch man who comes from the netherworld.***
>
> ***A khakhua possesses the body of a man (it can never be a woman) and begins to magically eat their insides, according to***

> ***logic of the Melanesian imperative you must pay back in kind. They must eat the khakhua as it ate the person who died. It is part of their revenge based justice system.***

He was introduced to a native called Kili-Kili, the most notorious khakhua killer of them all. He has killed and eaten twenty-three possessed men. Raffaele was handed the skull of the latest victim, a man who had once been a good friend.

> ***They had chopped off the top of the skull to get at the brains—their favourite. They steam everything with an oven made from leaves and rocks. They treat it like they would the flesh of a pig. They cut off the legs separately and wrap them in banana leaves. They cut off the head and that goes to the person who found the khakhua. That's why Kili-Kili had the skull. They cut off the right arm and the right ribs as one piece and the left as another. I asked them what it tasted like, and although you always get this common misconception that it tastes like pig they say the flesh tastes more like Cassowary—a New Guinea and Northern Australian bird that resembles an ostrich or an emu.***
>
> ***They eat everything except the hair, nails, and the penis. Children under 13 are not allowed to eat it, because they believe that as they are eating the khakhua it is very dangerous—there are evil spirits all around and the children are too vulnerable.***

The Korowai do not consider somebody possessed by a khakhua to be human any longer. When they eat them, they are eating the khakhua, not the person. Apparently, according to Kornelius, the killing and eating of khakhua still goes on to this day.

In Fiji, cannibalism was still practised up to the early twentieth century due to the belief that one could absorb the victim's qualities. *The Guinness Book of Records* lists the world's most prolific cannibal as coming from Fiji. Ratu Udre Undre was a Fijian chief who, in the nineteenth century, is said to have eaten between 872 and 999 people. He made a row of stones, leaving a rock for every victim. Ratu Udre Undre was shot and killed by Fiji government officials in 1840.

Reverend Richard Lyth, who was staying at a place called Cokova in northeastern Viti Levu in 1848, was taken to see the late Udre Undre's cannibal stones.

> ***Ravatu, a son of the above prince of cannibals took me out of the town about a mile to show me the stones by which his father memorialised the number of human beings that he ate from time to time beginning after his family had begun to grow up, or as Ravatu also expressed it when he was beginning to be a little gray.***
>
> ***Those that he ate in his youth and up to the period referred to were included. I was brought to a long line of stones placed close together in a row. They lay a few paces from the path and alongside it.***
>
> ***Ravatu assured me that his father ate all this number of human beings. He added a stone to the row for each one he received.***

> ***They were victims killed in war. He ate them all himself, he gave to none. However much he had on hand, it was cooked and re-cooked (by which it was preserved) until it was all consumed. He would keep it in a box so he would lose none. Turtles and prime pigs he would reserve for himself but at the second time of cooking he would give a portion to his children and friends but not so human flesh. He kept it entirely for himself. He ate but little else very little vegetable and being an enormous eater, he was able to get through a great deal.***

In the 1890s, five slave girls on Malaita in the Solomon Islands were slaughtered and eaten in honour of the visit of the French count, Rodolphe Festetics de Tolna. The count took their photo shortly before they were killed. He did not state if he had eaten the meat of the slaughtered girls.

Hans Stefan von Kotze (1869–1909), the German travel writer, visited New Ireland, largest island in the Bismarck Archipelago. He was invited to a cannibal feast. The main course was a young woman Kotze himself had given money for the feast! He did not see the girl slaughtered but observed her being cooked in an earth oven and on spits. He felt to sickened to eat any of her flesh.

Moving on to New Zealand, the French privateer, Marc-Joseph Marion du Fresne and his crew anchored his ship at the Bay of Islands in May of 1772. At first the men were on good terms with the native Māori, until they began to sneak abroad and steal items from his ship at night.

The French had a hospital tent on the shore, as many of them were suffering from scurvy. They were raising plants in a cultivated area for vitamins. Māoris were seen creeping around the camp at night. In the morning, they returned with a gift of fish. The French demonstrated their guns to the natives. On the afternoon of June 12, 1772, Marion and a group of armed sailors went to the Māori village. Marion and twenty-six men of his crew were killed.

One survivor, who had escaped, told his shipmates that the Māori had tricked them into going into the bush, where they had been ambushed, with all the others being killed. That night, four hundred armed Māori suddenly attacked the hospital camp but were stopped in their tracks by the fire of multiple blunderbusses.

Next day, 1,200 Māori attacked. Some had on the clothes of their victims. The French decided to abandon the hospital camp and returned to the ship. The Māori then stole all the tools and supplies and burnt the camp down.

They returned the next day, heavily armed. 1,500 natives attacked but were set to flight by the French guns. The French chased them to their village, killing 250 of them. The rest escaped in canoes. Later they found a sailor's cooked head on a spike, as well as human bones near a fire. The Māori had eaten the sailors.

Later, in December of 1809, the British brigantine the *Boyd* anchored at Whangaroa Bay in the Northland peninsula in New Zealand. The vessel was there to pick up wooden spars made from local kauri trees. The ship was under the command of Captain John Thompson and carried about seventy people. Among them was Te Ara, son of a Māori chief who had been working on a number of ships for about a year. On board the *Boyd*, he had been falsely accused of stealing pewter spoons and whipped with a cat-o-nine-tails.

Te Ara regained the confidence of the captain and persuaded him to put into Whangaroa Bay, assuring him that it was the best place to secure the timber he desired. Whangaroa, however, was where Te Ara's tribe lived. Once there, Te Ara ran off to his people and reported his ill treatment. The tribe swore revenge.

The Māori invited Captain Thompson and his chief officer, and three others followed the canoes to where they could find the best kauri trees. The rest of the crew stayed aboard with the passengers, preparing the vessel for the voyage to Britain. Once out of sight, the natives killed the sailors with clubs and axes and took their clothes. The bodies were taken back to their village to be eaten. The Māori disguised themselves as the sailors and returned to the ship. In the night, they boarded the *Boyd* and stealthily crept about it, killing the crew and dismembering them.

Five people were hidden in the rigging and survived. Next morning, they saw a large canoe belonging to another chief, Te Pahi, who had come to trade with the Whangaroa people. He rescued the survivors, but two Whangaroa canoes pursued them. All but one of the survivors were killed.

Five other people on board the ship had been spared. These were Ann Morley and her baby, apprentice Thomas Davis, the second mate, and two-year-old Elizabeth Isabella Throsby, who was taken by a local chief and kept for three weeks before being rescued. The little girl, whose mother had been killed and eaten by the Māoris, was taken back to Sydney, Australia, on a whaling boat and reunited with her father. She went on to marry, have many children, and live a long life.

The *Boyd* itself was towed towards shore by the natives but got caught on mudflats. The Māoris ransacked it, taking all the goods, but accidentally set off a keg of gunpowder that exploded, killing ten of them and setting the ship ablaze.

The survivors from the *Boyd* were rescued by a ship called the *City of Edinburgh*, whose crew found the remains of the cannibal feast strewn on the beach. Captain Berry captured two Māori chiefs responsible for the massacre, at first holding them for ransom for the return of survivors.

In the 1830s, a European trader named Anscow saw how a fifteen-year-old slave girl was killed with a tomahawk in a Māori village, apparently as punishment for having been absent without permission. Her body was then dismembered and the flesh washed in the river, cooked, and eaten. The next day, when Anscow moved on, some villagers insisted on accompanying him, carrying small baskets of the girl's cooked flesh, which they brought as gifts to a nearby village.

EUROPE

Cormac O Grada, the Irish economic historian, found references to cannibalism in the great famine of 698–700. He also records cannibalism for a famine in 1116 and for several in the sixteenth and seventeenth centuries, including reports of little children being killed so they could be eaten.

Italy has seen more than its fair share of cannibalism in the past.

In 1305, there was an episode connected to the suspected murder of John I, the last Marquis of Montferrat, a descendant of the House of Aleramici, and the only male heir of William VII of Montferrat. The episode is narrated in the *De Gestis Civium Astensium* by Guglielmo Ventura.

John fell seriously ill after the surrender of the city of Asti in January 1305. A few days later, he died heirless, after having entrusted the management of his land to the commune of Pavia. His personal physician, Maestro Emanuele of Vercelli, was charged with his sudden death. The accusation, according to Guglielmo, was unfounded. Yet, as soon as the funeral rites were performed, the ministers of the deceased marquis murdered Emanuele of Vercelli without a trial, stabbing him to death, and many among them ate his flesh.

Six years later, in 1311, Henry VII of Luxembourg, after crossing the Alps at the head of his army, was besieging the rebel city of Brescia. The Brescians defended themselves bravely and roasted and devoured their slain enemies. The imperial soldiers acted in kind when they captured Tebaldo Brusato, the leader of the resistance; they wrapped him in a cowhide and dragged him around the city walls, then they beheaded and quartered him, and exposed his remains at the four corners of the city. The besieged, in revenge for the outrage to the noble Brusato, captured a nephew of the emperor, who they roasted and ate.

In 1343, in Florence, a revolt broke out against Walter of Brienne, duke of Athens and lord of the city. On July 26, the insurgents barricaded Brienne in his palace, forcing his followers to flee. To appease the fury of the rebels, the duke handed them Guglielmo di Assisi, a public administrator, and his son. The two unfortunate men were killed on the spot and dismembered into small pieces. Some insurgents brandished pikes with shreds of their spoils, while others, more daring, devoured their flesh.

In 1368 in Montepulciano, Giacomo del Pecora, tyrant of the city, the only despot remaining after the expulsion of his brother Niccolo, barricaded himself inside the city under siege. On February 4, the besiegers penetrated the city with the support of the people of Montepulciano and captured Giacomo. The chronicler Donato di Neri narrates that the following day, the people of Montepulciano freed the tyrant only to kill him: the insurgents cut him to pieces and ate his limbs.

On Sicily in 1377, the city of Geraci was disputed between the Ventimiglia and Chiaromonte families. Following a new dispute with the Palizzi, allies of the Chiaromonte and supported by the new sovereign Pietro II, Francesco Ventimiglia had been branded a traitor and besieged in Geraci. When Francesco died, murdered or perhaps killed in an accident while attempting to escape, the inhabitants of Geraci sheared his fingers, gouged his eyes out, pulled his teeth off by hitting them with a stone and cut his beard off with the flesh. His body was then cut into pieces, and some ate his liver.

On September 3, 1385, the people of Ferrara, exasperated by heavy taxation, revolted against Tommaso da Tortona, who was responsible for the tax policy. Tommaso was beaten with whips and sticks, hit with blades, wounded with hooks, stoned, and cut with axes, then dragged from the square to the stake on which the rioters had burned the records of duties and taxes. There, they extracted his liver and heart to devour them. Other remains were hung on pikes and sticks and paraded through the city. Some of the gruesome trophies were exposed in plain sight at the port as a warning. In the end, the little that remained of the body was set on fire together with books and documents.

In the village of Acquapendente, in Central Italy in 1437, a man murdered a boy who had accidentally killed his son and served the boy's cooked body parts to his father in an act of revenge. This led to a feud between the two families, which took the lives of thirty-six people during the following month.

The Italians seemed to settle down for a bit, but then it all kicked off again in Milan in the year 1476. Gian Galeazzo Maria Sforza, the Duke of Milan, was a cruel man. He once had a poacher executed by forcing him to swallow an entire hare (with fur intact), had another man nailed alive to his coffin, and a priest who had predicted a short reign was punished by being starved to death. This made him many enemies in Milan. It was also said of Sforza that he had raped the wives and daughters of numerous Milanese nobles, that he took sadistic pleasure in devising tortures for men who had offended him, and that he enjoyed pulling apart the limbs of his enemies with his own hands.

The duke was stabbed in the chest and groin by Giovanni Andrea Lampugnani and his accomplices while he was going to Saint Stephen mass. As the duke died, the murderers fled, but Lampugnani stumbled and was killed by the duke's coachman. He was then dragged through the city, where citizens ate his heart, hand, and liver.

In Forli in 1488, the bodies of the conspirators who murdered Girolamo Riario, lord of Imola and Forli, were also butchered. Leone Cobelli, a direct witness who sided with the lords of Forli, provided a detailed narration of the incident. Countess Caterina Sforza, once she regained control of the city, applied strong measures of justice against the persecutors of her late husband. The father of the Orsi brothers, heads of the conspiracy, was captured and killed in place of his fugitive sons: tied to a board with his head left outside, he was hung on the tail of a horse and dragged three times around the square. His corpse was then quartered, and his entrails scattered on the ground. Cobelli said that a soldier picked up his heart and ate it.

The infamous "Red Wedding" occurred on the night of July 14, 1500, at the marriage of Astorre Baglioni and Lavinia Colonna. Carlo di Oddo Baglioni had conceived, together with his brother-in-law Girolamo della Penna, the atrocious plot to exterminate all the members of his own family by killing Guido and Ridolfo Baglioni together with all their children. The conspirators invaded the palace after the festivities. Filippo di Braccio, together with various followers, was assigned to the wedding alcove, where he killed Astorre, still lying on the bridal bed. Filippo di Braccio then extracted the heart from his chest and ate it. He then left his naked body in the middle of the street. Giampaolo Baglioni, who had managed to escape and obtain external help, later overwhelmed the rebels and regained control of the city government. The incident inspired an episode of the fantasy show *Game of Thrones*.

A second event occurring in the same year in Acquasparta, near Todi, in which the same Giampaolo who had escaped the massacre of his family was involved. On July 16, 1500, an army under the orders of Vitellozzo Vitelli, Giampaolo Baglioni, and Giulio and Paolo Orsini, supported by the Pope, besieged Acquasparta to free Todi and the surrounding territories from the dominion of Altobello di Chiaravalle and Girolamo da Canale. When the city of Acquasparta was stormed, Altobello was intercepted while trying to escape. Along the way to the prison, an angry mob snatched the prisoner from the guards. A contemporary chronicle attributed to Francesco Maturanzio states that "Every man came to kill him, to the point that the executioners in their frenzy, hurt each other. The remains of the tyrant were devoured with such fury that nothing was left of his miserable and unfortunate body."

In 1501, the struggle between the opposing consortia of the Chancellors and the Panciatichi turned to the worse for the latter. After taking refuge in Serravalle, the Panciatichi were betrayed by some members of their faction and torn to pieces. Some had their hearts ripped out and bit to pieces. A pattern of heart eating is emerging in Italy!

There was a period of peace again until 1585. Following the decision of the Elected Representatives to increase the price of bread on May 9, Giovan Vincenzo Starace was dragged to a public assembly in the monastery of Sant Agostino, where his election had once taken place. Here, after a failed attempt to escape the rioters, he was attacked with sticks, stones, and knives. His body was then stripped, castrated, mutilated, and dragged upside down through the city streets. Thomas Costo, who lived at the time, wrote:

They all rushed over that unhappy body like wild beasts, and tore it into many pieces, someone cutting off a hand, someone a foot and a piece of a leg, someone the arms, someone the ears, someone the nose, someone a limb and others something else. They then ripped out his guts, the heart and the other limbs. When all the limbs were ripped into small pieces, they divided them and put them on top of sticks, on the tips of swords and other weapons.

The bloody relics were paraded around the city on spikes, with some eating them raw and others taking them home to cook.

The lesson here is, "Don't piss off the Italians!"

The great famine that swept Europe from 1315 to 1317 caused outbreaks of cannibalism. Harsh winters and wet, cold summers caused crop yields to plummet, livestock to die, and food prices to skyrocket. The climate change, known as the "Little Ice Age" may have been caused by volcanic activity. The human population of Europe dropped by 42 percent.

The chronicles of Bristol in England record that in 1315 there was:

A great Famine of Dearth with such mortality that the living could scarce suffice to bury the dead, horse flesh and dog's flesh was accounted good meat, and some ate their own children. The thieves that were in Prison did pluck and tear in pieces, such as were newly put into prison and devoured them half alive.

Other such accounts were recorded across Europe during these times.

CRIMINAL CANNIBALISM

Criminal cannibalism is usually a solo affair. An individual will kill and eat other humans for a number of reasons ranging from mental unbalance to strange sexual gratification. It is far beyond the scope of this book to record all instances of such behaviour, so we will look at some of the most infamous and the most weird. Animals kill for food, in territorial disputes, over breeding, or if they feel threatened. But man kills for sadism.

One of the strangest cases of cannibalism occurred in eighteenth-century France. Born in Lyon, France, in approximately 1772, this man is remembered only by his stage name, Tarrare. From a young age, he was cursed with an unnatural hunger, eating vast amounts of food. It was said that he could eat his own body weight in cow meat in twenty-four hours. His family could not afford to feed him, and he was forced to leave home.

Little is known about his early life on the road, but he joined a group of travelling performers. His tricks included swallowing small animals and inanimate objects whole and eating fantastic amounts of food. His appearance was strange. He had a wide, thin-lipped mouth, stained teeth, and despite his inhuman gluttony, he was as thin as a rake, his belly hanging down like a deflated balloon. When engorged with food, his belly would look like that of a pregnant woman. He had bloodshot eyes, was constantly sweating, and had a foul body odour that others found intolerable.

Tarrare was conscripted into the French Revolutionary Army. Here he garnered the interest of Dr. Pierre-Françis Percy. Tarrare could not subsist on army food, even when given the rations of four men. He took to scavenging around dung heaps but passed out from exhaustion. He was sent to the army hospital, where Dr. Percy met him.

It was found that he could eat as much as fifteen normal men and would also eat small animals, including an eel, an owl, lizards, snakes, and cats.

Tarrare was commissioned as a spy for the French Army of the Rhine, during the French War of the First Coalition. His amazing gut was to be put to good use. His first mission was to secretly courier a document across enemy lines in a place where it could not easily be detected if caught: his digestive tract. He was rewarded with a wheelbarrow full of bull innards that he apparently ate in front of the disgusted generals.

He was given a small wooden box containing a message for an important prisoner of war in Prussia. Tarrare was to pass it through his system and give it to the prisoner. Unfortunately, Tarrare did not speak German and was quickly caught and beaten, and narrowly avoided execution.

Incapable of continuing his military service and desperate to find a cure for his condition, he was once again under the care of Dr. Percy. Laudanum opiates, wine vinegar, tobacco pills, and a diet of soft-boiled eggs were all tried, to no avail. At night Tarrare walked the streets, looking for remains from slaughterhouses. Back at the hospital he was found, multiple times, devouring dead bodies. Eventually a toddler vanished, and it was thought that Tarrare had killed and completely eaten the child. A mob chased him from the hospital and the city.

Some four years later, Dr. Percy was contacted by a physician of Versailles hospital at the behest of a patient on their deathbed. It was Tarrare, who told him he had swallowed a golden fork two years before and it had lodged inside him. Dr. Percy found that it was not the fork but tuberculosis that was killing him.

After his death a postmortem was carried out. Tarrare possessed a shockingly wide oesophagus which allowed spectators to look directly from his open mouth into his stomach, which was unfathomably large and lined with ulcers. His body was full of pus, his liver and gallbladder were abnormally large, and the gold fork was never recovered.

Tarrare may have had an enlarged hypothalamus. The hypothalamus regulates the body's temperature and is responsible for causing the sensation of hunger. This may explain Tarrare's sweating and ravenous hunger. He may also have had worms from eating raw meat. Tarrare was by no means the most prolific cannibal, but he was surly one of the weirdest!

Antoine Leger was a cave-dwelling hermit who lived in a cave in Charbonniere Rock, located above Moutmiraux, France. On August 12, 1824, he saw twelve-year-old Aimee Constance Debully walking by. He strangled her with his handkerchief, carried the body to the woods, and had sex with it. He then ate several parts of the body, including the girl's heart. He took her mangled corpse back to his cave and buried it in a shallow grave in the sand.

A large search was undertaken for the missing girl. Leger's handkerchief was found and on August 16, the entrance to his secret cave was found. The stench of the rotting body was overpowering. Leger was arrested and his trial took place on November 23.

Leger tried to plead insanity but was guillotined on November 30.

In his 1865 book *The Book of Were-Wolves, Being an Account of a Terrible Superstition*, the British clergyman Sabine Baring-Gould recorded a case that occurred in 1849 in Polomia, southern Poland. A vagrant named Swiatek murdered a fourteen-year-old girl and ate her body. Swiatek also admitted to having killed and eaten five other people since 1846, although evidence was found of up to fourteen victims. He claimed that he had developed a taste for human flesh three years previously after hunger obliged him to eat the body of a man killed in a tavern fire.

In 1917, a horrific case emerged from Iraq. The country had been experiencing a famine due to inflation. Abboud and Khajawa were a couple living in Mosul, where Abboud was a tour guide and Khajawa was a cook. During the famine the couple and their young son had subsisted on the meat of stray dogs and cats, but in 1917 they strangled an elderly woman who had come to visit them. They cooked the corpse and tried to eat it but found the old woman's flesh too tough and fatty. Khajawa suggested that they try eating children instead.

They got their young son to encourage children to come and play at their house, whereupon they would bludgeon the child to death with a rock. The body would be skinned, dismembered, cooked, and eaten. The victim's skulls were thrown down a well in their backyard.

Abboud and Khajawa continued this for several months, killing dozens of children, often from the same families. Due to the ongoing famine, the children were seldom missed, and the authorities knew nothing. Abboud and Khajawa even opened a small restaurant, selling the meat of murdered children as mutton.

They were undone when one man who brought meat from them found a small bone in the portion he was eating. Being a butcher, he recognized it as a human finger-bone. He reported it to the police, who inspected the home of Abboud and Khajawa, finding the collection of skulls in the well. The couple were arrested.

After a short trial, they were sentenced to death. On the morning of their execution, the couple rode two donkeys to Bab Al-Tob Square, where makeshift gallows had been erected. On the way, many bystanders cursed, spat at and assaulted the couple, with one report claiming that a woman whose three children had been killed by them bit off one of Khajawa's toes. They were hung in front of a large crowd of onlookers.

Albert Fish was born in Washington, DC, in 1870, into a family with a history of mental illness. In 1875 his father died of a heart attack and his mother sent Albert to Saint John's Orphanage in Washington, where he was physically abused and grew to enjoy the pain. In 1880 Fish's mother secured a job and brought him back from the orphanage aged twelve, but the abuse and latent mental illness had already begun to create a monster. He began a homosexual relationship with a telegraph boy of the same age. The pair would drink each other's urine and eat each other's faeces. He also began writing obscene letters to women whose names he acquired from matrimonial agencies and classified ads. A grown man doing these things would be cause for alarm, but Fish was not even a teenager.

At twenty, he moved to New York and became a male prostitute. He also began raping little boys under six. Fish's mother arranged a marriage for him to Anna Mary Hoffman in 1898. They had six children, but family life did not put an end to Fish's perversions. He was later convicted of grand larceny and sent to Sing-Sing maximum security prison.

After his release, he took a male lover who took him to a wax museum where he saw a model of a bisected human penis. After that he became obsessed with genital mutilation. In 1910, whilst working in Wilmington, Delaware, Fish began a sadomasochistic relationship with an intellectually disabled nineteen-year-old, Thomas Bedden. Ten days into the relationship, Fish took Bedden to an abandoned farmhouse, tied him up, and tortured him for the next two weeks. He finally cut half his penis off, saying, "I shall never forget his scream, or the look he gave me."

Fish poured peroxide over the man's member, wrapped in in a Vaseline-covered hanky and left, taking the first train home. "I never heard what become of him, or tried to find out."

Fish's wife left him for the lodger John Straube, leaving him with the children. He never abused his own children, but he did encourage them and their friends to paddle his buttocks with a nail-studded paddle. He also began inserting needles into his groin and abdomen and dousing cotton wool with lighter fluid, inserting it into his anus, and setting it on fire!

On July 14, 1924, Fish claimed his second victim, nine-year-old Francis McDonnell. Francis's mother had seen Fish shambling down Decker Avenue at Port Richmond, Staten Island, New York.

> ***He came shuffling down the street mumbling to himself and making queer motions with his hands... I saw his thick grey hair and his drooping grey moustache. Everything about him seemed faded and grey.***

Her description would later lead to Fish being called "The Grey Man."

Francis had been playing a ball game when he was snatched. Fish sexually assaulted him, then strangled him with his own suspenders. The boy was found hanging from a tree. Flesh had been stripped from his left hamstring.

On February 11, 1927, three-year-old Billy Beaton and his twelve-year-old brother were playing in the apartment hallway in Brooklyn. Four-year-old Billy Gaffney joined them, and later the older boy left. Both the younger boys vanished. Billy Beaton was later found on the roof of the apartments. Beaton said that the "bogeyman" had taken his friend. The bogey man was Albert Fish. The child's body was never found, but in a subsequent, stomach-churning letter to his attorney, he describes the little boy's fate. This is not for the faint-hearted.

> ***I brought him to the Riker Ave. dumps. There is a house that stands alone, not far from where I took him... I took the G boy there. Stripped him naked and tied his hands and feet and gagged him with a piece of dirty rag I picked out of the dump. Then I burned his clothes. Threw his shoes in the dump. Then I walked back and took trolley to 59 St. at 2:00 a.m. and walked home from there. Next day about 2:00 p.m., I took tools, a good heavy cat-of-nine tails. Home made. Short handle. Cut one of my belts in half, slit this half in six strips about 8 in. long. I whipped his bare behind till the blood ran from his legs. I cut off his ears, nose, slit his mouth from ear to ear. Gouged out his eyes. He was dead then. I stuck the knife in his belly and held my mouth to his body and drank his blood. I picked up four old potato sacks and gathered a pile of stones. Then I cut him up. I had a grip with me. I put his nose, ears and a few slices of his belly in the grip. Then I cut him thru the middle of his body. Just below his belly button. Then thru his legs about 2 in. below his behind. I put this in my grip with a lot of paper. I***

cut off the head, feet, arms, hands and the legs below the knee. This I put in sacks weighed with stones, tied the ends and threw them into the pools of slimy water you will see all along the road going to North Beach. Water is 3 to 4 ft. deep. They sank at once. I came home with my meat. I had the front of his body I liked best. His monkey and pee wees and a nice little fat behind to roast in the oven and eat. I made a stew out of his ears, nose, pieces of his face and belly. I put onions, carrots, turnips, celery, salt and pepper. It was good. Then I split the cheeks of his behind open, cut off his monkey and pee wees and washed them first. I put strips of bacon on each cheek of his behind and put in the oven. Then I picked 4 onions and when meat had roasted about 1/4 hr., I poured about a pint of water over it for gravy and put in the onions. At frequent intervals I basted his behind with a wooden spoon. So the meat would be nice and juicy. In about 2 hr., it was nice and brown, cooked thru. I never ate any roast turkey that tasted half as good as his sweet fat little behind did. I ate every bit of the meat in about four days. His little monkey was as sweet as a nut, but his pee-wees I could not chew. Threw them in the toilet.

The *New York World* newspaper ran an advert on May 25, 1928, "Young man, eighteen, wishes position in country. Edward Budd, 406 West 15th Street." Fish visited the Budd family and introduced himself as "Frank Howard." His plan was to mutilate and kill Edward Budd. Fish told the man he would contact him in a few days. When he returned, Budd's ten-year-old sister Grace was at the house. Fish now fixated on her and quickly made up a story about having to attend his niece's birthday party. Fish invited Grace to the party and her mother and father said yes, a fatal mistake.

The monster took Grace to an abandoned house, a cottage at 359 Mountain Road, East Ervington. There Fish strangled Grace, decapitated her, and ate her corpse over several days.

The police arrested the wrong man, sixty-six-year-old superintendent Charles Edward Pope, on September 5, 1930, as a suspect in Grace's disappearance, accused by Pope's estranged wife. The unfortunate man spent 108 days in jail before being found not guilty.

In the meantime, the bogey man, the grey man, had escaped.

In November 1934, Fish sent an anonymous letter to Grace's mother and father. Again, be warned—this is the stuff of nightmares.

My dear Mrs. Budd,

In 1894 a friend of mine shipped as a deckhand on the steamer Tacoma, Capt John Davis. They sailed from San Francisco to Hong Kong China. On arriving there he and two others went ashore and got drunk. When they returned the boat was gone. At that time there was a famine in China. Meat of any kind was from 1 to 3 Dollars a pound. So great was the suffering among the very poor that all children under 12 were sold to the Butchers to be cut up and sold for food in order to keep others from starving. A boy or girl under 14 was not safe in the street. You could go in any shop and ask for steak, chops, or stew meat. Part of the naked body of a boy or girl would be brought out and just what you wanted cut from it. A boy or girls behind which is the sweetest part of the body and sold as veal cutlet brought the highest price. John staid there so long

he acquired a taste for human flesh. On his return to N.Y. he stole two boys one 7 one 11. Took them to his home stripped them naked tied them in a closet then burned everything they had on. Several times every day and night he spanked them, tortured them, to make their meat good and tender. First he killed the 11 yr old boy, because he had the fattest ass and of course the most meat on it. Every part of his body was cooked and eaten except Head, bones and guts. He was roasted in the oven, (all of his ass) boiled, broiled, fried, stewed. The little boy was next, went the same way. At that time I was living at 409 E 100 St, rear, right side. He told me so often how good human flesh was I made up my mind to taste it. On Sunday June the 3, 1928, I called on you at 406 W 15 St. Brought you pot cheese, strawberries. We had lunch. Grace sat in my lap and kissed me. I made up my mind to eat her, on the pretence of taking her to a party. You said Yes she could go. I took her to an empty house in Westchester I had already picked out. When we got there, I told her to remain outside. She picked wild flowers. I went upstairs and stripped all my clothes off. I knew if I did not I would get her blood on them. When all was ready I went to the window and called her. Then I hid in a closet until she was in the room. When she saw me all naked she began to cry and tried to run down stairs. I grabbed her and she said she would tell her mama. First I stripped her naked. How she did kick, bite and scratch. I choked her to death then cut her in small

pieces so I could take my meat to my rooms, cook and eat it.

How sweet and tender her little ass was roasted in the oven. It

took me 9 days to eat her entire body. I did not fuck her, though,

I could of had I wished. She died a virgin.

The letter was Fish's undoing. The hexagonal emblem that was on the letter was traced to the New York Private Chauffeur's Benevolent Association. A caretaker for the association had taken some stationary home to his boarding house at 200 East 52nd Street and left it there when he moved. Fish had been living at the apartment. The police swooped and arrested him.

The trial lasted ten days. James Dempsey, the defence counsel, said that Fish was a "psychiatric phenomenon" and that nowhere in legal or medical records was there another individual who possessed so many sexual abnormalities. Fish was found guilty and was executed on January 16, 1936, in the electric chair at Sing-Sing. Thus ended the bogey man.

Fish's lawyer Dempsey revealed that he was in possession of his client's "final statement." This amounted to several pages of hand-written notes that Fish had apparently written just prior to his death. When pressed by the assembled journalists to reveal the document's contents, Dempsey refused, stating, "I will never show it to anyone. It was the most filthy string of obscenities that I have ever read."

Fish was also suspected of killing and mutilating seven other children.

I personally find cases like this much, much, more disturbing that animal attacks. A crocodile, bear, or shark may kill and eat you out of hunger. The grey man killed for sadism and sexual satisfaction. If there is such a place as hell, I hope he screams in it forever.

Karl Denke, unlike Albert Fish, was a well-liked member of the community in the town of Oberkunzendorf, Prussia (now part of Poland). Born in 1860, he became a gardener and a cross bearer and organist for the Lutheran church. Locals called him "papa." He opened a shop selling leather goods and meat.

Denke began to kill homeless drifters. Nobody knows why he did this. He killed Ida Launer in 1903. Six years later, in 1909, he killed twenty-five-year-old Emma Sander. Slaughterhouse worker Eduard Trautmann was found guilty of Sander's murder and imprisoned until 1926. This allowed Denke to carry on his work for years. He kept a ledger of his killings and sold the meat in his shop, as well as goods made from human skin. He ate the meat himself as well and pickled some to sell in the nearby village of Breslau. He told they buyers that it was pork.

On December 21, 1924, Denke lured a homeless drifter named Vincenz Olivier into his home. Denke attacked him with a pickaxe, but Oliver escaped with only a cut to the head. He tried to alert the authorities but was initially disbelieved due to Denke's good reputation. Oliver was instead arrested himself for vagrancy. The judge, however, insisted on further investigation of Olivier's claims, whereupon Denke was taken in for questioning. Whilst in his cell, Denke managed to hang himself. A police search of his house found boxes of human bones, pots of human fat and meat, and the cannibal's murder ledger that had the names of thirty victims in it.

Alonzo Robinson didn't do things by halves. He was a grave robber, murderer, and cannibal. From an impoverished family in Cleveland, Mississippi, Robinson was arrested in 1918 at the age of twenty-three for sending obscene letters to women but escaped on the way to jail. Moving to Ferndale, Michigan, he took the nom-de-plume of "James H. Coyner."

In 1927, he stole a girl's corpse from her grave. He ended up imprisoned at the Indiana State Penitentiary in Michigan City. Whilst there he told the prison authorities, "I could tell you a lot of things you'd like to know, a lot of things about murders and murderers, but I'll die first."

Bodies of decapitated women were turning up around Ferndale and Robinson was a suspect, but the killings could not be linked to him at the time. After being paroled in June 1934, he returned to Cleveland and began writing dirty letters again. In December, at the home of Aurelius B. Turner and his pregnant wife, Mr. Turner, who was reading a newspaper, was hacked with an axe, then shot. His wife was struck five times with the axe. He mutilated the bodies and took away some of Mrs. Turner's flesh.

Robinson was arrested on January 12, 1935. Police found salted human flesh in his pockets with bite marks on it. A trunk found in the area had the missing women's heads in it. The trunk belonged to Robinson. The freak was hung in March of 1935, grinning as the noose was put around his neck.

Leonarda Cianciulli lived in the province of Avellino. In 1917 she married Raffaele Pansardi, a registry office clerk in Lacedonia. After their home was destroyed in the war, they moved once more to Correggio, where Cianciulli opened a small shop. She was very popular and well respected within her neighbourhood.

The couple had seventeen children, but three were miscarried and ten more died in their youth. She became very protective of the remaining four. She had been informed by a fortune-teller that she would have many children, but all would die young. A Romany palm reader had also told her, "In your right hand I see prison, in your left a criminal asylum."

In 1939, with Europe on the brink of war, her eldest son Giuseppe was going to enlist in the Royal Italian Army. Fearing for his life, she came to the insane conclusion that only human sacrifice could save him. She set about killing three women to protect her son.

Faustina Setti was a spinster who had come to Cianciulli hoping the woman could help her find a husband. Cianciulli said she knew of a suitable man in the province of Pola. She convinced Setti not to tell anybody before she left and to write letters to friends and relatives to be mailed when she reached Pola, to tell them that everything was fine. Setti gave Cianciulli her life savings of 30,000 lire for her help. When she visited Cianciulli again before she left, the mad woman killed her with an axe, dragged her body into a closet, hacked it apart, and caught the blood in a basin.

I threw the pieces into a pot, added seven kilos of caustic soda, which I had bought to make soap, and stirred the mixture until

the pieces dissolved in a thick, dark mush that I poured into several buckets and emptied in a nearby septic tank. As for the blood in the basin, I waited until it had coagulated, dried it in the oven, ground it and mixed it with flour, sugar, chocolate, milk and eggs, as well as a bit of margarine, kneading all the ingredients together. I made lots of crunchy tea cakes and served them to the ladies who came to visit, though Giuseppe and I also ate them.

Cianciulli's second victim was Francesca Soavi, who was looking for work. Cianciulli claimed to have found her a position in school for girls in Piacenza. As with her last victim, she convinced Soavi not to tell anybody where she was going. Cianciulli gave her drugged wine and killed her with an axe. The body was disposed of the same way as before. Cianciulli also obtained 3,000 lire from Soavi.

Virginia Cacioppo was a former soprano. Cianciulli claimed to have found her work as the secretary for an impresario in Florence. Once more, the victim was told not to tell anybody about her new position. The murder was the same, but this time Cianciulli used the woman's body to make soap.

She ended up in the pot, like the other two... Her flesh was fat and white, when it had melted I added a bottle of cologne, and after a long time on the boil I was able to make some most acceptable creamy soap. I gave bars to neighbours and acquaintances. The cakes, too, were better: that woman was really sweet.

From Cacioppo, Cianciulli reportedly received 50,000 lire, assorted jewels, and public bonds. She sold her victim's clothing and shoes.

Albertina Fanti, Cacioppo's sister-in-law, grew suspicious and reported her disappearance to the police. As Cacioppo was last seen entering Cianciulli's house, the police arrested her. Cianciulli did not confess to the murders until their suspicion fell on her son, Giuseppe Pansardi. She then provided detailed accounts of what she had done to absolve her son from any blame.

She was sentenced to thirty years in prison and died in 1970 of a stroke.

After WWII, in the Australian War Crimes Section of the Tokyo Tribunal, the leaders of Imperial Japan were tried for war crimes. The prosecutor, William Webb, collected instances of cannibalism by Japanese soldiers on their captives. As the Japanese soldiers' rations grew more meagre, the instances of cannibalism grew more frequent. Historian Yuki Tanaka says, "cannibalism was often a systematic activity conducted by whole squads and under the command of officers."

Indian prisoner of war Lance Naik witnessed Japanese cannibalism in New Guinea.

> ***The Japanese started selecting prisoners and every day one prisoner was taken out and killed and eaten by the soldiers. I personally saw this happen and about 100 prisoners were eaten at this place by the Japanese. The remainder of us were taken to another spot 50 miles (80 kilometres) away where ten prisoners died of sickness. At this place, the Japanese again started selecting prisoners to eat. Those selected were taken to a hut where their flesh was cut from their bodies while they were alive and they were thrown into a ditch where they later died.***

In another case in late 1944, nine American pilots escaped from their planes after being shot down during bombing raids on Chichijima, largest of the Bonin Islands, a subtropical archipelago belonging to Japan. Eight of the airmen—Lloyd Woellhof, Grady York, James "Jimmy" Dye, Glenn Frazier Jr., Marvell "Marve" Mershon, Floyd Hall, Warren Earl Vaughn, and Warren Hindenlang—were captured and eventually executed. Japanese soldiers ate five of the American airmen. The ninth pilot, and the only one to evade capture, was future US President George W. Bush! Five Japanese soldiers were hung for their crimes.

Joachim Georg Kroll was a rapist, murderer, and cannibal who operated in the Ruhr region of Germany. At large for twenty-one years, he killed thirteen girls and women and one man. A cunning killer, he would wait months or years between victims. As other killers were operating in Germany, he went unsuspected for years.

A toilet attendant by trade, his modus operandi was to strangle his victims, strip them, and have sex with the corpse, then dismember and sometimes eat them.

In 1955 he stabbed, raped, and disembowelled Irmgard Strehl, aged nineteen, in Ludinghausen. A year later in Bottrop, he raped and strangled twelve-year-old Erika Schuletter.

Kroll left it until 1959 to strike again, when he killed Klara Frieda Tesmer, twenty-four. Heinrich Ott, a mechanic, was wrongly arrested for the killing and hung himself in jail. Kroll then raped and strangled Manuela Knodt, sixteen, in Essen. He carved flesh from her buttocks and thighs.

Waiting until 1962, he struck again in Dinslaken, raping and strangling thirteen-year-old Petra Giese. Another innocent man, Vinzenz Kuehn, was arrested and convicted. Two months later, he killed Monika Tafel, aged twelve, in Walsum. Once more, he cut meat from the victim's buttocks. Walter Quicker was arrested for the crime. He was later acquitted, but harassment from his neighbours drove the man to suicide.

1962 was Kroll's busiest year, as he struck again in Burscheid, abducting Barbara Bruder, twelve, whose body was never found.

Kroll laid low for three years before taking up his dark work again in August of 1965. Hermann Schmitz, twenty-five, and his girlfriend Marion

Veen were sitting in a car in a "lover's lane" in Duisburg when the maniac attacked. Schmitz fought Kroll as Venn escaped. The murderer killed her boyfriend, his only male victim.

In September of 1966, Kroll travelled to Foersterbusch Park near the town of Marl, where he strangled twenty-year-old Ursula Rohling. Her boyfriend, Adolf Schickel, was falsely accused of the crime and committed suicide. Next month, in the city of Wuppertal, Ilona Harke, aged five, was raped and drowned in a ditch.

The killer waited another three years before killing again. He raped and strangled his oldest victim, sixty-one-year-old Maria Hettgen, in Huckeswagen.

In 1970, Kroll strangled Jutta Rahn, thirteen, as she was walking home from a train station. Peter Schay was arrested and eventually released.

Kroll took his longest break after killing Rahn. It was not until 1976 that he killed again. In May of that year, he strangled and raped ten-year-old Karin Toepfer in Voerde. In July of the same year, he abducted and killed four-year-old Marion Ketter. As the police were going from house to house looking for the girl, a neighbour approached a policeman and told him that the wastepipe in his apartment building had blocked up, and when he had asked his neighbour, Kroll, whether he knew what had been blocking the pipe, Kroll had simply replied, "Guts." Hearing this, the police went up to Kroll's apartment and found the body of Marion Ketter hacked apart. Some parts were in the fridge, a small hand was cooking in a pan of boiling water, and the entrails were found stuck in the wastepipe. Kroll was immediately arrested.

Kroll stated that he ate the flesh of his victims to save on groceries! In custody, he believed that he was going to get a simple operation to cure him of his homicidal urges and would then be released from prison. Instead, he was given life imprisonment and died there in 1991.

Josef Kulik was born in 1930 in what was then Czechoslovakia. By sixteen, he was a criminal and was sexually abused by his own father. He served ten years for attempted murder. Whilst inside, he had sex with other inmates. On his release, he joined the Czechoslovakia People's Army.

On July 29, 1963, Kulik was supposed to return from leave to his unit in the town of Prelouc. Kulik fell asleep in the train and missed his stop. The train was now in the town of Pardubice, which was the end of the line. Walking back, he reached Rosice, a suburb of Pardubice. Here he broke into a parked station wagon.

A little while later, Vladimir Drtina, six, and Oldrich Krenek, nine, entered the wagon to play. Kulik murdered the boys with an axe and a knife. He removed some of the entrails from their bodies and then masturbated over them. He took a kidney and spleen from Vladimir and Oldrich's heart and kidney and handkerchief. He then stole some old wreaths from a nearby graveyard and made a fire. He impaled the entrails on a sharpened stick, roasted them, and ate them. He then fled the scene.

A police search for the boys with helicopters was launched that night. It was thought that the boys had drowned in the River Elbe. On August 2, a bike matching the one ridden by one of the missing boys was discovered in a marsh near a railway embankment. While inspecting the surroundings, the investigative team noticed a smell that was spreading near one of the parked railcars. After breaking down the door, the bodies of the two boys were discovered. A large number of footprints were found at the site.

Meanwhile, Kulik was walking cross-country for home, the village of Litomerick. On August 2, Kulik fell asleep in a field near the village of Brnikov. He was run over by a tractor that crushed his legs. Seriously injured, he was taken to a hospital in Rosice. Among his belongings was a notebook in which he had stupidly written down his deeds. He was arrested in hospital.

As he was a soldier of the socialist army, the investigation was covered up, and the entire trial of Josef Kulik was closed to the public. He was hung February 7, 1964.

Richard Trenton Chase of Sacramento, California, showed serious behavioural problems from an early age. He believed that his cranial bones had separated and were moving about. He shaved his head to observe this. He would hold oranges to his head in the belief that he could absorb vitamin C through diffusion. He also thought that somebody had magically stolen an artery from his heart.

In 1973, he was put into a mental institution after injecting rabbit's blood into his veins. Whilst in hospital, he caught and killed birds and drank their blood. He was diagnosed with paranoid schizophrenia. After treatment, he was released in 1976 and returned to live with his mother. Later, he finally got his own apartment.

In 1977, he was arrested when found near Pyramid Lake, Nevada, covered with blood and holding a bucket of blood. The blood was found to be cow's blood and he was released.

On December 29, 1977, Chase killed fifty-one-year-old Ambrose Griffin in a drive-by shooting. Then, on January 23, 1978, he broke into the house of pregnant Teresa Wallin and shot her. He then raped her corpse and stabbed it. He removed multiple organs, cut off one of her nipples, and drank her blood. Finally, he took dog faeces from his victim's backyard and forced them down the body's throat.

Five days later, Chase broke into the house of thirty-eight-year-old Evelyn Miroth. First, he shot her friend Danny Meredith. Then he shot Miroth, her six-year-old son Jason, and her twenty-two-month-old nephew, David Ferreira. He then had sex with Miroth's body and cannibalised it.

When a neighbour knocked on the door, Chase fled in Miroth's car, taking Ferreira's body with him. The police were called, and they found handprints and shoe imprints in Miroth's blood.

Chase was arrested in his apartment where it was found that the walls, floor, ceiling, refrigerator, and all of Chase's eating and drinking utensils were soaked in blood.

Despite pleading insanity, a jury found Chase guilty of first-degree murder on May 8, 1979, and sentenced him to die in the gas chamber.

FBI agent Robert Kenneth Ressler was granted interviews with Chase. The killer told him he was forced to kill by Nazis to stay alive! He had a deep fear of Nazis and UFOs, and asked Ressler to get him a radar gun to apprehend the Nazi flying saucers so that the Nazis could stand trial for the murders they forced him to do.

On December 26, 1980, he committed suicide by an overdose of prescription drugs.

Nikolai Dzhumagaliev had the nickname "iron fang." He had lost his front teeth in a fistfight and had them replaced with metal ones. Dzhumagaliev was born to a Kazakh father and a Belarussian mother in Uzun-Agach, Kazakhstan, in 1952. His childhood was normal, and he did national service in the army. Then in 1973 he applied to go university. He failed to get a place, and shortly after failed to secure a job as a chauffeur. Dzhumagaliev instead went trekking through the Urals, Siberia, and Murmansk. Returning home in 1977, he got a job as a fireman.

Soon after his return, he was found to have sexually transmitted diseases, syphilis and trichomoniasis. Dzhumagaliev blamed women as a whole, and he developed an intense hatred for them. Two years later he claimed his first victim.

He stalked a woman walking alone by the side of the road between Uzun-Agach and Maibulak. Dragging her off the road, he slashed her throat and drank the blood. He undressed the body and cut out the breasts, organs, hips, and thighs, which he put in a backpack and took home. He cooked the flesh and ate it for the next month. A murder investigation was launched, but was later abandoned after no leads were found. Over the next six months, Dzhumagaliev killed and ate five more women.

During a drunken fight with another fireman on August 21, Dzhumagaliev shot and killed his colleague. He was arrested for manslaughter but found not guilty by reason of insanity after he was diagnosed with schizophrenia at Moscow's Serbsky Institute.

Released after one year, Dzhumagaliev returned home and killed another three women. He served their meat to unwitting friends who came to visit him. In one such event, he murdered a female friend with an axe. As he did this, two other friends were in the room next door. Horrified at what they saw, they fled and contacted the police. Dzhumagaliev did not realize he had been seen.

When the officers came to his house, they found the fiend naked, covered in blood, and still hacking at the body. They were so shocked that the madman escaped. He was arrested the next day in his cousin's home. While in custody, Dzhumagaliev confessed to the women's murders, claiming that they were prostitutes and that he wanted to rid the world of them. He was tried on December 3, 1981, and was once again found insane. This time, the court established that he should be sent to a mental clinic where he would receive compulsory treatment.

On August 29, 1989, whilst being transported to another madhouse, Dzhumagaliev escaped and fled to the mountains of Kyrgyzstan. He hid out in the forest, collecting medicinal plants that he traded for food.

Dzhumagaliev contacted a friend and convinced him to send a letter to his family from Moscow, to make the government believe that Dzhumagaliev was still in the capital. The letter finished with his reassurance that he would not return home because there were many women in Moscow and nobody would miss them. This information was picked up by the newspaper *Kurants*, which also reported that Dzhumagaliev had been seen in the city and its surrounding region.

By April of 1991, Dzhumagaliev was tired of his feral existence in the mountains. He devised a plan to get arrested for a petty crime and sent to jail under a false identity. He stole sheep in Fergana, Uzbekistan, with the aim of being imprisoned in the Uzbek capital, Tashkent. After his arrest, he confessed to the theft and claimed to be a Chinese citizen. He failed to explain what he was doing in the Soviet Union, and the police got suspicious and requested help from Moscow. As a result, Moscow detective Yuri Dubyagin travelled to Fergana, and he immediately recognized Dzhumagaliev. He was sent to an asylum again.

After the fall of the USSR, Dzhumagaliev was declared sane again and was to be sent home. But there was an outcry from the residents of Uzun-Agach. Eventually, he was moved to a high-security mental clinic in Aktas, a village near Almaty, where he remains today and is allowed to work as a repairman. Dzhumagaliev petitioned unsuccessfully to be given the death penalty during his third institutionalization. In 2014, he was charged with the 1990 murder of a female student in Aktobe, western Kazakhstan, whose death fit Dzhumagaliev's MO.

One of the sickest and most prolific American serial killers was Jeffery Dahmer. Born in Milwaukee in 1960, he went on to kill seventeen victims between 1978 and 1991. As a boy, he became obsessed with animal bones and collecting dead animals to "see how they worked." In later life, Dahmer would pick up young men, drugging them, then killing them by strangulation or by drilling a hole into their skull into which he would inject boiling water or hydrochloric acid. He would keep body parts such as heads and skulls as trophies of his murderous acts and have sex with the corpses. It is beyond the scope of this book to examine all of his crimes, so we will instead concentrate on the victims whom he cannibalised. These he killed whilst living in his grandmother's apartment in West Alllis, Wisconsin.

Raymond Smith was a thirty-two-year-old male prostitute whom Dahmer met in 1990. He lured him back home with the promise of fifty dollars for sex. His victim was given a drink laced with sleeping tablets before being strangled. Dahmer photographed the body in suggestive positions, then dismembered it in a bathtub. He cut off and consumed the heart, liver, biceps, and portions of thigh, using condiments to flavour them and tenderizing the meat beforehand. He boiled the rest of the body before dissolving it in acid. He kept Smith's skull, which he spray-painted and kept in a cabinet.

In September of the same year, he brought nineteen-year-old Ernest Miller back to his apartment. He gave him coffee, again laced with sleeping tablets, but only had two pills to hand. When the drink didn't have the full effect, Dahmer slashed his throat. He photographed the man's body naked and dissected it in the bath. He talked to and kissed Miler's severed head. He cut out Miller's heart, liver, and biceps and placed them in the freezer. He later ate them. He then bleached his victim's bones, defleshed the head and retained the skull.

On July 15 of the following year, Dahmer offered to pay twenty-four-year-old Oliver Lacy to pose for nude pictures. The pair had sex, and

Dahmer attempted to prolong the time he spent with his victim alive by using chloroform. When this failed, he returned to his old MO and drugged and strangled Lacy. He dismembered Lacy and kept his skeleton, heart, and head in the freezer. He ate some of the man's flesh again.

Six days later, the killer made a mistake. Thirty-two-year-old Tracy Edwards had gone home with Dahmer to drink beer and pose for photos. Dahmer handcuffed Edwards and produced a knife, telling the man that he intended to eat his heart. Edwards convinced the madman to put the knife away and release him by promising to pose nude for pictures. Dahmer agreed, and afterwards, Edwards escaped by punching him in the face and running out of the house.

When Edwards told the police, two officers found photographs of dead and dismembered men at Dahmer's apartment. Later, the Milwaukee police's Criminal Investigation Bureau did a search of Dahmer's playpen of evil, unearthing four severed heads in the kitchen, seven skulls in the bedroom, two hearts in the fridge, and an entire torso, plus a bag of human organs in the freezer. Elsewhere in Apartment 213, investigators discovered two entire skeletons, a pair of severed hands, two severed and preserved penises, a mummified scalp, and, in a fifty-seven-gallon drum, three further dismembered torsos dissolving in an acid solution. A total of seventy-four Polaroid pictures detailing the dismemberment of Dahmer's victims were also found.

Dahmer was given life in prison and sent to Columbia Correctional Institution where he eventually got his comeuppance. On November 28, 1994, fellow inmate Christopher Scarver attacked Dahmer with a metal rod whilst they were on work duty together. They had been cleaning the gym toilet, and Scarver removed the rod from a piece of gym equipment and battered the cannibal to death.

You may think that Jeffery Dahmer was the last word in sadistic, horrific crime, but you would be wrong, Andrei Chikatilo, the Butcher of Rostov, makes him look like an amateur. Between 1978 and 1990, he sadistically killed and mutilated fifty-seven people!

Chikatilo was born on October 16, 1936, in the village of Yabluchne in the former USSR. He grew up in a background of war, poverty, and hunger. As a young man, Chikatilo found that he suffered from erectile dysfunction and was unable to satisfy any woman he was with. Many mocked him and gossiped about his impotence. He soon developed a burning hatred for women.

His family arranged a marriage for him in 1963 to Feodosia Odnacheva. Despite being unable to have sex in the normal way, he fathered a son and daughter by manually inserting his semen into his wife with his hands after masturbation. Apparently, he treated his wife and children well and held down a job as a teacher for ten years before being sacked due to his molestation of a number of pupils. He later got a job as a clerk for a factory based in Rostov which produced construction materials. The job allowed him to travel across much of the Soviet Union to buy or order materials.

In 1978, he brought a dilapidated hut in a run-down area of Rostov. In September of that year, he lured nine-year-old Yelena Zakotnova back there and attempted to rape her. Unable to get an erection, he stabbed and strangled the girl to death, ejaculating as he did so. Thus, a killer was born. Finding sexual release in sadism and murder, Chikatilo targeted women—prostitutes, runaways, and the homeless. He brutally stabbed, slashed, hacked, and dismembered them whilst getting sexual satisfaction. Many times, he cut out his victims' uteruses and ate them. He began to target young boys too, killing them and hacking off their genitals in envy of his own not working.

Over the next twelve years, he butchered fifty-seven victims. The Soviet police were unable to catch him due to their out-of-date methods and the arrogant belief that a serial killer could not operate under communism. They theorized that a satanic cult was at work, or that organs were being harvested for sale on the black market. Eventually, they changed tack and started to use criminal profiling methods perfected in the US. Consulting psychiatrist Dr. Alexandr Bukhanovsky was brought in and crime scene and medical examiner's reports were made available to him so he could produce a psychological profile.

Dr. Bukhanovsky concluded that the killer was between forty-five and fifty, had had a painful, isolated childhood, and was incapable of courtship with women. He was well educated, probably married, was a sadist, and was impotent. His only way of sexual release was through murder. As the victims had been killed on weekdays and the bodies found near transport hubs, then the killer must travel a lot in his work. Based upon the actual days of the week when the killings had occurred, the killer was most likely tied to a production schedule. All in all, a startlingly accurate profile.

Finally, police put Chikatilo under surveillance, as he fitted the profile. He was seen trying to pick up young women around train stations. He was finally arrested after being seen emerging from woodland and washing blood from himself. He had a folding knife and two lengths of rope with him.

Chikatilo later confessed to the murders, thinking that he could fake insanity and be free of the death sentence. During his trial, he was kept in a specially constructed metal cage to stop the families of his victims attacking him. He was found to be sane, stood trial, and was found guilty and subsequently executed by a bullet to the back of the skull on February 4, 1994.

Daniel Paul Rakowitz was a marijuana dealer living in East Village of New York in the 1980s. He was known for having a pet rooster. He made up his own religion, called The Church on 966, and convinced himself that the homeless people who he talked to in Tompkins Park Square were his followers.

In 1989, he began bragging to them that he had murdered his girlfriend, Monika Beerle. Nobody had believed him. But he had, in fact killed the woman and dismembered her in a bathtub. He had boiled the body parts, and served some of her remains in the form of a soup to the very homeless who had doubted him. He had eaten soup made from her brain and liked it. Some of the homeless told the police, who arrested Rakowitz.

He took them to a storage facility at the Port Authority Bus Terminal in Manhattan, where he had kept Beetle's skull and teeth.

Deemed insane, he was incarcerated in Kirby Forensic Psychiatric Center, where he remains to this day.

Nathaniel Bar-Jonah was a morbidly obese child molester. He lived in the Montana town of Great Falls, where he had moved from Massachusetts, where he had just finished a long sentence for the sexual assault and attempted murder of a young boy.

Bar-Jonah's story is a strange one. In 1964, he was given a Ouija board for his seventh birthday. He tried out the board in his cellar with a five-year-old neighbour girl. He tried to strangle the girl, but fortunately his mother heard her screaming and rescued her.

In 1970, when he was thirteen, he lured a six-year-old boy into going sledding with him. He then sexually abused him. In 1975, claiming to be a policeman, he lured an eight-year-old boy into his car and sexually assaulted him, then tried to strangle him. Luckily, a neighbour saw what was happening and phoned the police. Bar-Jonah only received a year's probation.

Three years later, he abducted two boys from a movie theatre and drove them to a remote area to abuse them. To silence them, he locked one boy in the boot of his car and strangled the other, leaving him for dead, then drove away. Fortunately, the boy survived and ran to get the police, who arrested Bar-Jonah with the other boy still in the boot of his car. He was convicted of attempted murder and sentenced to twenty years in prison.

Later, a prison psychologist listened to his fantasies about abusing, killing, and eating children. The psychologist recommended that he be moved to a mental asylum.

Incredibly, in 1991, he was deemed no longer be a danger and was released to live with his mother in Montana. Just days after being released, Bar-Jonah spotted a seven-year-old boy sitting in a parked car. He forced his way into the car and tried to smother the boy by sitting on him. He was stopped by the boy's mother and arrested by the police. Nobody from the

Massachusetts court followed up with the probation officers in Montana, to which Bar-Jonah had fled.

Back in Montana, he was soon luring boys to his apartment to abuse them, and even erected a pulley for hanging them. At the same time, he was making food for his neighbours out of meat none of them could recognize. Bar-Jonah claimed it was venison, but nobody ever saw him go hunting.

In 1999, he was arrested outside a local elementary school, carrying a fake gun and dressed as a police officer. On searching his house, police found thousands of photos of children cut from magazines and a weird journal written in some kind of code. They also found a human bone.

By the time the trial began, the FBI had decoded Bar-Jonah's journal. Inside, he described his obsession with torturing and murdering children. There was also a list of twenty-two names. Eight of them were known to be Bar-Jonah's earlier victims. Many of the rest were local children. The others were never identified. The book also included his plans to cook and eat children.

A ten-year-old boy, Zachary Ramsay, had recently vanished, and it was thought that Bar-Jonah had killed him, put his body through a meat grinder, and eaten him, giving portions to his neighbours.

Bar-Jonah was jailed for 130 years but died of heart disease in 2008.

Joshua Milton Blahyi was a general in the brutal Liberian civil war, which lasted from 1989 to 1997. He was known as General Butt-Naked, as he would run naked into battle with a machete in the belief that his nakedness gave him protection.

In 2008, he confessed before a Truth and Reconciliation Commission to the murders of thousands of people, claiming that during the war, he had magical powers due to his initiation into a traditional secretive society when he was eleven years old. He claimed to have dragged children underwater and broken their necks.

Every time a city was taken, I had to make a human sacrifice to maintain my power. They would bring me a living child and I would cut out its heart, which I would eat.

He would cut out the hearts of babies and share them with his followers.

Blahyi later converted to Christianity, repented, and became a pastor!

In Russia, Ilshat Kuzikov killed and ate three men. Born in Tajikistan in 1960, he saw his father murder his mother, and later claimed that afterwards he had lost all compassion. After being raised by his aunt, he moved to Saint Petersburg. He was married for a time but soon divorced.

Kuzikov began hanging around with homeless people, inviting them back to his apartment to drink vodka. In 1995, locals began to find body parts in rubbish. Investigations led to Kuzikov, who left blood in the stairwell outside his apartment, which neighbours complained was emitting a rotten smell. In his apartment, police discovered dried human skin and buckets of cooked human flesh.

As it turned out, he had killed, cooked, and eaten three men over the course of a year. Investigators thought that, in actuality, he had killed and devoured far more. Kuzikov claimed to have killed them purely for food, as he could not live on his poor wages. He was sentenced to compulsory treatment at a psychiatric hospital along the Arsenalnaya embankment in Saint Petersburg. Some sources suggest he died in the early 2000s.

Russia seems to have more than its fair share of cannibals. Mikhail Yuryevich Malyshev was born on October 31, 1965, in Perm, Russia. His family was well-to-do and well respected. However, his parents divorced, affecting him mentally and emotionally. He was also bullied at school for being overweight. At this time, he began to kill cats and dogs. In 1980, he was arrested on charges of cruelty against animals and interned at a juvenile detention centre.

He took up boxing and decided to eat more meat to build up muscle mass. There was a shortage of meat products at the time, so he resorted to eating dog meat. As he became more aggressive, his stepfather and mother banned him from their home.

His mother got him an apartment in which he lived and continued to kill stray dogs. He took in an eighteen-year-old mentally ill orphan named Nikolai, whom he sexually abused. He sent the youth out to catch stray dogs that Malyshev killed and ate. He began drinking heavily at this point.

The following year, he met a nineteen-year-old nurse named Inna Borovi, who moved in with him. Soon he was raping and torturing both her and Nikolai and forcing them to have sex with each other. He threatened to throw them into the freezer if they tried to escape. He turned the apartment into an insanitary brothel with rubbish and animal carcasses strewn about the place.

In 1997, he sent out Borovi to find him a girl to rape. She convinced sixteen-year-old Natalia Suvorova to come back to the apartment and drink with her. Once there, Malyshev raped the girl, and when she threatened to tell the police, he hacked her to death with an axe. He forced Borovik and Nikolai to dismember the body. Borovik had a nervous breakdown during the act, and Malyshev savagely beat both her and Nikolai to force them to comply. He also bit off and ate the tip of Borovik's nose. Malyshev then put Suvorova's hands, feet, and head in a bag, which he then threw into the River Kama.

The victim's remains were found, but an innocent resident of Perm was arrested and coerced into a confession. Nobody suspected Malyshev.

A year later, Malyshev invited a man named Anton to drink with him at the apartment. After a night of boozing, Malyshev killed him with an axe. He dismembered the corpse and ate his heart. Making no attempt to hide his crimes, Malyshev threw the body into the rubbish behind his house.

After identifying the victim, the police questioned his brother, who stated that shortly before his disappearance, Anton had gone to visit an acquaintance named Mikhail Malyshev. Searching the apartment, the police found packages of human meat.

Borovik and Nikolai decided to testify against Malyshev, who soon cracked and confessed. He told the police that he ate human meat because he wanted to know how it tasted and he thought it would build up his muscle mass.

After his conviction, Malyshev was transferred to the high-security penal colony in Chusovoy. He stayed out of trouble in prison, and after twenty-three years, was released and moved back to the apartment in which he had carried out his crime, much to the outrage of locals.

Known as "El Comegente," the people-eater, Jose Dorangel Vargas Gomez was a man-eating tramp from Venezuela. Born in 1957 to a poor farming family, Gomez had twice been arrested for stealing cows and chickens. In 1995, he decided to up the ante by killing Baltazar Cruz Moreno and eating his carcass. Gomez was locked up in the Institute of Psychiatric Rehabilitation Peribeca. What happened next beggars belief.

As he struck the doctors as being "normal" (remember, he had just killed and eaten a man), he was released! Then, between November 1998 and January 1999, the cannibal tramp killed and ate another eleven men. He was living rough in a park in San Cristobel and hunted his victims with a tube-shaped spear and rocks. His targets were mainly athletes training in the park and labourers working on the riverbank. He only ate men, explaining later that he thought that women and children were "too pure" to kill. How very noble of him. As his shack had no fridge to keep meat cool, he ate what he could of his victims quickly, then dumped the remains. This also meant that he had to kill quite often.

On February 12, 1999, members of the civil defence found the remains of two bodies. Soon after, six more bodies were found. At first it was thought they had been killed by a drug cartel or a satanic cult. When some of the remains were identified, it was found that they were on a missing persons list.

Police searched Gomez's shack and found several vessels containing human flesh and viscera prepared for consumption, along with three human heads and several feet and hands. He was taken to prison for good this time. In 2016, during a prison riot, he killed two inmates and served them to other prisoners.

Back in the bastion of urban cannibalism, Russia, Igor Viktorovich Churasov, or "the Scavenger of Humanity," killed seven people between 1997 and 2000, eating a number of them. Churasov was born in 1966 into a normal, happy family in Rayann Oblast and lived a normal life at school and work, showing no tendencies to violence or crime. In the 1990s, he joined a circus as an assistant, uniformer, prop designer, and animal care worker.

It was at this time that he fell in with ex-convict Gennady Emelkin. They rented a wooden house in the Gorbaty Bridge area. Soon the first signs of mental illness awoke in Churasov. He became convinced that humanity had no right to life and should be exterminated. Emelkin agreed with him, and the pair decided to start murdering.

Between 1997 and 2000, Churasov committed seven murders, Emelkin assisting in five of them. They lured victims back to their home, then strangled, stabbed, or bludgeoned them to death, dismembered them, and burnt the body parts in an old stove.

On one occasion, Churasov was aided by a third accomplice, twenty-three-year-old Natalia Makartsova, who lured Albina Noskova, also twenty-three, back to the men's house knowing full well that they would kill her. Churasov strangled her with a hose, then cut out her heart and liver and fried and ate them. From then on, he always kept the meat and organs of his victims and ate them.

In the spring of 2000, Churasov and Shurmanov were arrested after a friend saw them murdering a man. They admitted to the killing, and later led police to a ravine where they dumped their victims' body parts. According to investigator Igor Kurkin, Churasov considered himself a "sanitary of society who freed it from unnecessary garbage, fallen people." He also recalled one time when he found a skull in the Lazarevskoye Cemetery, brought it home, washed and polished it, then poured some earth into it and started growing strawberries. Churasov claimed that he considered the strawberries to be a representation of life, while the skull represented death.

Diagnosed with a volatile form of schizophrenia, Churasov was locked away in Sychevsk Special Hospital. Shurmanov was interred in the same facility. Makartsova was convicted and sentenced to six years' imprisonment.

There cannot be many cases where the victim volunteered to be killed and eaten of their own free will. But in our next case, that was just what happened. Armin Meiwes was born on December 1, 1961, in Germany. He claimed that his interest in cannibalism started with the child-eating witch in the fairy story Hansel and Gretel.

In 2001, he turned his fantasy into a reality. Advertising online, he asked to meet "a well-built eighteen- to twenty-five-year-old to be slaughtered and then consumed." A few men met up with Meiwes but then backed out. Eventually, Bernd-Jugen Armando Brandes, a forty-three-year-old engineer from Berlin, got in touch and visited Meiwes's home in the small town of Wustefeld. What followed was recorded on a videotape that thankfully has never been shown.

Dosing Brandes up with twenty sleeping pills and a bottle of cough syrup, Meiwes amputated his willing victim's member and they both tried to eat it. Meiwes fried it in a pan with wine, garlic, and pepper, using Brandes's fat. However, the penis became too burnt to eat.

Meiwes ran his victim a bath, then later killed him by stabbing him in the neck. He dismembered the corpse and then froze it, consuming the body over the next ten months. Meiwes was arrested in December 2002, when a college student alerted authorities to new advertisements for victims online.

Initially charged with manslaughter, Meiwes was jailed for eight years and six months. In a later retrial, he was jailed for life, as it was clear this was a premeditated murder for sexual gratification. In prison, he became a vegetarian!

Yoo Young-chul was a Korean serial killer and cannibal born in 1970 in Gochang County. Married with a child, Young-chul nonetheless killed twenty people and partially ate many of them. He began in November of 2003 by breaking into the homes of senior citizens and bludgeoning them to death. Police were confused as no money was taken. The killings were done for sadistic pleasure.

In January of 2004, he began targeting prostitutes and masseuses. He would call them to his home, have sex with them, then bludgeon them to death, dismembering the corpses and eating their livers. The bodies were disposed of in the surrounding mountains.

On July 15, 2004, Young-chul raised suspicions by calling a massage parlour where several employees had recently gone missing after receiving similar phone calls, so the owner of the massage parlour, accompanied by several employees and a single police officer, went to the agreed-upon meeting place. The police officer left before Yoo arrived, and Yoo was apprehended by the employees of the massage parlour. Another police officer placed handcuffs on Yoo after he was detained by the massage parlour employee.

Faking an epileptic fit, he escaped from the police station, but was recaptured twelve hours later. He was sentenced to death in December of 2004.

Wouldn't you know it, we are back in Russia. Fifteen-year-old Konstantin Baranov founded a Satanist gang in Yaroslavl in 2006. According to the investigation, he was joined by other teenagers, Nikolai Ogolobyak, Alexey Chistyakov, Anton Makovkin, Sergey Karpenko, Alexander Voronov, and Ksenia Kovaleva. Together they performed bloody rituals, sacrificing dogs and cats.

In the summer of 2008, they killed four teenagers, Olga Pukhova, Anna Gorokhova, Varvara Kuzmina, and Andrey Solovyov. According to the investigation, they were stabbed to death and their bodies were dismembered. The gang then recited several incantations over their bodies, scalped them, and cut off their genitals. If this were not enough, the bodies were beheaded and the tongues, hearts, and breasts were eaten. To cap it all, the Satanists then had necrophilic sex with the mangled corpses.

In July 2010, a Yaroslavl court sentenced the cultists to prison for terms ranging from two to twenty years. Nikolai Ogolobyak had been a legal adult at the time and got the longest sentence. He was sent to a high-security penal colony, having been found guilty of the murder of two or more people, and the desecration of the bodies of the dead. He was set to be released in 2030. All the others who were sent to prison have now been released.

In 2023, Ogolobyak was pardoned after having volunteered to fight in Vladimir Putin's evil war on Ukraine. He fought in the Storm Z Unit, which consisted mainly of criminals, until he was badly injured and forced to stand down.

On July 30, 2008, carnival barker Tim McLean was riding home to Winnipeg in a Greyhound bus after working at a fair in Edmonton. When the bus stopped at Erickson, Vince Li, a man of Chinese descent but living in Canada, got on. He settled next to McLean, who had fallen asleep. Li then produced a large knife and killed McLean by stabbing him in the neck and chest. The driver pulled the bus over, and the other passengers fled. Attempts to get to McLean's body were thwarted when Li slashed at people with his knife.

Li decapitated his victim and displayed the severed head to the other passengers outside before beginning to eat it. For several hours, Li fed on McLean's flesh. At eight thirty that evening, the Royal Canadian Mounted Police arrived after being alerted. Other passengers had prevented the cannibal madman's escape. The bus driver and a truck driver had provided a crowbar and a hammer as weapons.

A stand-off followed, with a negotiator and an armed tactical unit. All the time, Li continued to feed on his victim's body. Li attempted to escape by driving the bus away, but the driver had engaged the emergency immobiliser unit. Finally, the freak tried to escape through a window and was captured after twice being blasted with a taser. The victim's ear, nose, and tongue were found in Li's pockets.

Li had serious mental problems, with voices telling him that he was the "third story of the Bible" and the "second coming of Jesus." They also informed him that he was destined to save the earth from an alien invasion. The voice would regularly order Li to travel through the country, on foot or by bus, often disappearing from his home for days on end. He thought that alien infiltrators were constantly after him. His wife said that he often went several days without sleeping or eating, and often spent his days at home crying and telling her about his visions of God. She and many of Li's friends attempted to get Li to visit a doctor for his auditory hallucinations, but he refused due to a fear of hospitals. They finally separated. Li was diagnosed with schizophrenia by staff and refused medication.

Judged unfit to stand trial, he was incarcerated in the Selkirk Mental Health Centre. He was discharged in 2017.

Surprise, surprise, we are in Russia again. Alexander Vladimirovich Bychkov of Belinsky targeted elderly and often homeless men. He was born to alcoholic parents in 1988; his father committed suicide by hanging. Alexander had to leave college to take care of his brother, who had brain damage after being beaten and thrown out of a car.

On September 17, 2009, Bychkov killed sixty-year-old Yevgeni Zhidkov. The man was heading to Belinsky district archive to fill out the documents needed for drawing a pension. He met Bychkov n a pub and the younger man offered him a place to sleep at his house. When Zhidkov fell asleep, Bychkov stabbed him to death, dismembered the body, and ate his liver, heart, and some muscle.

In the coming years, Bychkov killed more elderly men, mostly homeless alcoholics, whom he killed with a knife and hammer, feeding on the bodies before getting rid of the leftovers on a rubbish dump. He killed mainly in summer so that the police would suspect migrant workers.

In the spring of 2010 the first corpse was found, which belonged to Sergei Berezovsky, an ex-partner of Alexander's mother, Irina Bychkova. In September of the same year, two more dismembered corpses were found. Alexander Zhuplov, a local mentally ill man, was arrested for the murders on September 19, 2010. Zhuplov confessed to all three of the murders, and he was found guilty of all three and sent to an asylum.

On January 12, 2012, Bychkova broke into a hardware store and stole knives and money. Three days later, he was arrested. Police found a diary at his house that detailed his killings and cannibalism. He had killed and eaten eleven men. Found mentally competent to stand trial, he was sentenced to life imprisonment.

Stephen Griffiths of Bradford, England, began his life of crime at seventeen. Whilst on a shoplifting spree, he attacked a supermarket manager with a knife. He was sentenced to three years in prison. Whilst behind bars, Griffiths said he fantasised about becoming a serial killer. Five years later, he had another stint in the clink after holding a knife to a girl's throat.

Upon his release, he studied criminology at Bradford University. During that time, he killed, dismembered, and ate three sex workers, forty-three-year-old Susan Rushworth, Shelley Armitage, and Suzanne Blamires, thirty-six. The women were killed between June 2009 and May 2010. Parts of Suzanne's body were found in the River Aire in Shipley. Other human tissue found there was later discovered to belong to Shelley. Suzanne's remains were never recovered.

Griffiths killed the women at his flat with a crossbow and dismembered them in his bath before dumping the remains that he did not eat into the river in black plastic bags. He had kept Suzanne alive for two days before killing her. She had tried to escape, but he recaptured her. This was caught on CCTV. Footage showed Griffiths returning to the camera and flicking his middle finger up to it while holding up the crossbow. It was the caretaker at the flats who informed police of the distressing clips, and Griffiths was arrested.

Griffiths appeared in court, giving his name as "The Crossbow Cannibal." He also boasted to detectives that he had eaten some of his victims before chopping them up.

He was given a life sentence and will never be released.

THE CAUSES OF CANNIBALISM

In culture-bound cannibalism, endocannibalism is the eating of somebody from the same culture, mainly in funerary rites such as those in New Guinea that led to the laughing disease. It is thought to help souls cross over and is part of the grieving process, however alien it may seem to us.

Exocannibalism, on the other hand, is when somebody from another tribe is eaten. This is done as a celebration of victory over an enemy. It is also a common belief that by eating a victim, the eater will gain the eaten's powers or characteristics.

Related to this is human predation, where tribes go out and deliberately hunt, kill, and eat members of other tribes.

Medicinal cannibalism is the eating of human flesh in the belief that it is a kind of medicine and the ingestion of it will bring health benefits. This was common in China.

Infanticidal cannibalism is when children who are unwanted or thought unfit are eaten, usually by the parents. This was practised by Aborigines in Australia.

Sacrificial cannibalism is when human victims are killed for religious reasons and the flesh is consumed, mostly by priests. The Aztecs did this, as did the Celts in ancient Britain and the Bronze Age people in Knossos, Crete.

In some cases, like those in the Congo, humans are eaten because they taste good and are cheaper than pigs and goats. Slaves captured in battle would be bred and raised like livestock to be slaughtered and eaten. This is known as gastronomic cannibalism.

Criminal cannibalism is harder to define, and the dark, labyrinthine recesses of the human brain would take a whole volume to cover. Criminal cannibalism is thought of as an indicator of severe personality disorder or psychosis. Cannibalism may arise in thankfully rare cases of sexual fantasies that are acted out to their grim conclusions. A criminal cannibal may simply want to make their victim part of themselves forever, in the ultimate way. Jeffery Dahmer for example said that, by eating his victims, there was a "feeling of making him part of me."

Dr. Abbie Marono, Lecturer in Psychology at the University of Northampton, spent three years conducting research which seeks to understand the association between cannibalism and serial killers. She expected to find mental health concerns around the time of the kill playing a significant role, but that was not the case.

> ***Our results indicated that the factors which may differentiate cannibalistic serial killers from non-cannibalistic serial killers likely result from life history milestones rather than influences at the time of their killing. Evidence unearthed from this research suggests brain abnormalities, low social economic status, and abandonment as a child were some of the key factors for those who cannibalised their victims.***

Criminal cannibals, it seems, are made over a long period.

AVOIDING CANNIBALS

Most of us will never meet a cultural cannibal, unless we go to the most remote areas of New Guinea, Borneo, or tropical South America. Even in these regions, cultural cannibalism is vanishingly rare. Even the few cannibals left in these regions are unlikely to eat outsiders, least of all Westerners.

Criminal cannibals are another kettle of fish. Serial killers are rare, and cannibal serial killers rarer still. However, most of the criminal cannibals listed here were seemingly respectable members of society. Sure, some were tramps and eccentrics living on the outskirts of society, but most were not. Most presented like ordinary people. A cannibalistic murderer could be anybody, it could be your neighbour, your friend, a family member. How do you defend yourself against that? You can avoid the jungle, the tropical rivers, the sea, the deep woods, but you can't avoid the monsters that may lurk in our towns and cities. You may have passed one on the street today; you may have stood next to one at work. They look just like you and me.

END OF VOLUME ONE

And so, we have reached the end of volume one of *Creatures That Eat People*. The book was intended to be a single volume, but the subject matter is so vast that the publishers have decided to break it into two separate books. In volume two, we will look at big cats with a taste for man flesh, such as lions, tigers, leopards, and pumas, man-eating wolves and other wild dogs, killer hyenas, and unexpected man-eaters, such as primates, giant squid, the woman-eating elephant of World War II, and the host of crawling, buzzing, wriggling, slithering, twitching parasites that feed on our meat. I hope you can join me on part two of this grim but fascinating journey.

—RICHARD FREEMAN, EXETER, UK, 2024

ABOUT THE AUTHOR

Richard Freeman is a working cryptozoologist—he searches for and writes about unknown animals. The melodramatic may call him a monster hunter. He has hunted for creatures such as the yeti, the Mongolian death worm, the giant anaconda, the almasty, orang-pendek, the gul, the naga, the ninki-nanka, and the Tasmanian wolf. He is the Zoological Director at the Centre for Fortean Zoology, the world's only full-time mystery animal research organization, based in North Devon, England.

A former zookeeper, Richard has worked with over four hundred species—from spiders to elephants—but crocodiles are his favourite. He has lectured at the Natural History Museum in London, the Grant Museum of Zoology, and the Last Tuesday Society at Viktor Wynd's Little Shop of Horrors. Richard is also a regular contributor to the *Fortean Times*.

He has written several books about cryptozoology, folklore, monsters, and horror, including *Adventures in Cryptozoology*, *In Search of Real Monsters*, *The Great Yokai Encyclopedia*, and more.

Richard is a massive fan of classic *Doctor Who* ('60s/'70s, not the gender-flipped, woke, PC, modern rubbish) and a lover of weird fiction, horror, gothic music, and gothic ladies. Jon Pertwee and David Attenborough are his heroes. When not adventuring or searching for the unknown, he spends his days in Exeter, England.

Mango Publishing, established in 2014, publishes an eclectic list of books by diverse authors—both new and established voices—on topics ranging from business, personal growth, women's empowerment, LGBTQ studies, health, and spirituality to history, popular culture, time management, decluttering, lifestyle, mental wellness, aging, and sustainable living. We were named 2019 *and* 2020's #1 fastest growing independent publisher by *Publishers Weekly*. Our success is driven by our main goal, which is to publish high-quality books that will entertain readers as well as make a positive difference in their lives.

Our readers are our most important resource; we value your input, suggestions, and ideas. We'd love to hear from you—after all, we are publishing books for you!

Please stay in touch with us and follow us at:

Facebook: Mango Publishing

Twitter: @MangoPublishing

Instagram: @MangoPublishing

LinkedIn: Mango Publishing

Pinterest: Mango Publishing

Newsletter: mangopublishinggroup.com/newsletter

Join us on Mango's journey to reinvent publishing, one book at a time.